MICHIGAN BUSINESS
REPORTS   Number 58

Cable Television

Strategy for Penetrating Key Urban Markets

JAMES D. SCOTT

DIVISION OF RESEARCH   GRADUATE SCHOOL OF BUSINESS ADMINISTRATION

MICHIGAN BUSINESS
REPORTS   Number 58

able Television

rategy for Penetrating
ey Urban Markets

MES D. SCOTT

A publication of the
Division of Research
Graduate School of Business Administration
The University of Michigan
Ann Arbor, Michigan

Copyright © 1976
by
The University of Michigan

CONTENTS

TABLES

FIGURES

PREFACE

A virtual freeze on the expansion of cable television service into the top 100 television markets ended in March 1972, when Federal Communications Commission regulations were relaxed to provide a more favorable climate for the penetration of CATV into such areas. This report examines the key problems that must be overcome if this opportunity is to be exploited.

First, heavy investment is required to build or expand CATV systems in major metropolitan markets. Before such investment occurs, the prospect of profits must be favorable enough to justify the risks involved in making long-term commitments for expansion.

Second, if CATV revenues are to show satisfactory growth, system operators must develop and market a package of services that will attract new subscribers and get existing subscribers to buy added services. This task must be handled in a manner which will provide CATV with a competitive edge.

For the purposes of examining alternative possibilities, the major markets are classified by the degree to which existing over-the-air television meets the needs of their television viewers. The alternatives which CATV operators may consider in seeking a competitive edge to attract subscribers in the major markets include, among others: (1) importing the signals of distant independent television stations offering attractive programs; (2) offering premium entertainment (pay cable television); (3) originating programs which have special appeal for local television viewers; (4) interconnecting cable systems by microwave or satellite into program networks designed to make available to cooperating CATV operators those outstanding programs developed by other participants; (5) offering two-way interactive cable television services, such as home security systems and electronic shopping. This report discusses and evaluates these alternatives.

In planning his marketing strategy, the CATV operator needs to be consumer oriented. Market research is suggested as a means of estimating potential demand for service alternatives. Analysis suggests that marketing plans should be made in anticipation of the

xi

time when economic conditions improve. Execution of an aggressive, well-designed marketing strategy may then produce the desired payoff in market penetration.

Background and acknowledgments

This report was prepared as a part of a continuing research program dealing with the implications of the growth of cable television for marketing and advertising and the marketing problems of CATV system operators. The project is supported by the Sebastian S. Kresge Research Fund.

In the gathering of the data on which the report is based valuable assistance was rendered by Darrell Dahlman, Research Associate, who gathered data for Chapter 1 and conducted the studies reported in Chapters 5 and 6. Paul Hsu, Chairman, Department of Business Administration, National Chengchi University, Taiwan, also was of assistance, especially in connection with Chapter 1. Peter Robinson, Research Assistant, assumed responsibility for the study on pay-cable television (see Chapter 3). Several business executives, who must remain anonymous, also assisted by making available information which they had gathered on CATV. To all who contributed I express my sincere thanks.

I

BUILDING CABLE TELEVISION PENETRATION IN THE TOP 100 TV MARKETS

Introduction

Cable television (CATV) originally was developed to bring a clear television picture into homes which otherwise could receive either a poor signal or none at all. More recently it has been used to provide greater program variety to viewers in cities served by only two or three television stations. The advantages CATV offers to consumers have been great enough to lead them to pay installation fees ranging up to $100 (average $15) and monthly subscription fees averaging $5.40 as of 1974. CATV was at that time providing service to approximately 8.1 million subscribers (perhaps 25.92 million people), or about 12.5 percent of the total of U.S. television households. There were 3,100 operating cable systems serving 5,770 communities in the United States in 1974. Subscriber revenues had totalled $391 million in 1972.[1]

In examining the outlook for the future growth of CATV, it becomes apparent that much depends on the success experienced by CATV system operators in their efforts to increase subscriber penetration in the top 100 television markets. An A.C. Neilsen survey in 1969, for example, indicated that household penetration was 1.6 percent in the major metropolitan areas as compared with 34.5 percent in small towns and 23.3 percent in rural areas. This pattern of development was fostered by a number of economic factors to be discussed later, but it was also importantly influenced by a virtual freeze on CATV expansion into the top 100 television markets, which had resulted from regulations issued by the Federal Communications Commission.

These regulations were partially relaxed in 1972, improving the climate for CATV penetration into these 100 major markets and providing existing CATV operators in major metropolitan markets with an opportunity to expand subscriber penetration in their

1

service areas. Before penetration is likely to be expanded significantly, however, certain key problems must be overcome. It is our purpose to examine these problems, to note progress being made in their solution, and to outline future courses of action that appear to be necessary if the full potential of CATV is to be achieved.

Heavy investment is required to build or expand CATV systems in major metropolitan markets, where the cost of laying cable is estimated at $75,000 or more per mile. Thus providing the capital required for such expansion is an important problem to be solved, a task made more difficult by high interest rates, such as the 12 percent being charged on prime business loans as of August 1, 1974. The extent to which CATV revenues may be expected to grow in the top 100 markets is, therefore, significant in this regard.

A second problem in the penetration of major markets by CATV is developing and marketing a package of services which will attract new subscribers to the CATV system and encourage existing subscribers to buy additional services, thus increasing CATV system revenues.

A third challenge is the development of CATV program origination which will build subscriber audiences large enough to attract advertisers and thus provide essential revenues. Once such programs have attracted adequate audiences, the CATV system operator must convince advertisers (and their advertising agencies) that program audiences contain potential buyers of their products. Audience research, therefore, appears to be essential at this point. Once it has been completed it can be used in promotion directed to advertisers and their agencies to inform them of the opportunities CATV advertising offers.

Before these problems and opportunities are explored, however, it is desirable to view their character in proper perspective, i.e., against a background of information on the development of the CATV industry and on the outlook for its future.

Background

Development of CATV

Starting in 1949, CATV first developed in communities where, because of distance or topographical obstructions, there was no local television broadcasting station and reception from the nearest stations in the area was either nonexistent or poor.[2] The strong desire of people located in such communities for television entertainment led to the development of community antennae systems for receiving broadcast signals and feeding them through a network of coaxial cables to the homes of individual viewers on a subscription basis. Sensitive antennae were erected on a specially selected site; broadcast signals received were modulated, amplified, and fed to subscribers through a cable system.

Beginning in 1953 the original concept was supplemented by microwave relay systems which brought the broadcast signals of metropolitan-area stations over long distances to remote communities having little or no television service. This development then made it possible for cable-system operators to offer their subscribers both a clear television picture and a greater variety of programs than could be provided by the community antennae alone. Program diversity remains an important feature of present-day CATV.

According to *Broadcasting,* in 1974 most cable systems offered between eight and twelve channels; the average for all was ten; in practice they carried an average of seven signals. Stations constructed after March 31, 1972, are required to have 20 channels; by 1977, all systems must meet this standard. The state-of-the-art maximum is about 48. Technology exists for two-way cable television, which permits subscribers to transmit signals directly back to the originating cable operator. Effective March 1972, the Federal Communications Commission now requires the cable firms in the top 100 markets to have the capacity for such return communication, at least on a nonvoice basis.

Current status

CATV's growth has been one of the fastest in the communications field. Selected figures set forth in Table 1 demonstrate the progress that has been made since 1952 (as of January 1 of each year). Additional growth may be anticipated. Between 1972 and 1974 the number of operating systems increased from 2,750 to 3,100. These systems served 5,770 communities.

Although the average size of the systems operating in 1974 was approximately 2,400 subscribers, there were twenty-two systems with 20,000 subscribers or more. The largest—San Diego—served over 75,000. The greatest number of systems (805), however, fell in the 50 to 499 class, and there were thirty-eight with fewer than 50 subscribers each.

Of special significance is the percentage of TV homes served by CATV. Overall household pentration of CATV was 12.5 percent in 1974; this system had its greatest strength outside the major population centers, as indicated by the A.C. Neilsen 1969 figures on CATV penetration—23.3 percent of TV homes in rural areas,

Table 1

PROGRESS IN CATV, 1952 TO 1974

Item	1952	1962	1972	1974
Number of operating systems	70	800	2,750	3,100
Total subscribing households	14,000	850,000	6,000,000	8,100,000
Estimated number of viewers	*	*	18,500,000	29,920,000
Homes per system	200	1,062	2,182	2,400
Percentage of TV homes subscribing (household penetration)	.1	1.7	9.7	12.5

Sources: Broadcasting, *TV Fact Book*, No. 42, 1972–73; "Short Course on Cable, 1974," *Broadcasting*, Apr. 22, 1974, p. 23.

*Not available.

34.5 percent in small towns, and 1.6 percent of TV homes in major metropolitan areas. Why is CATV penetration so low in large population centers, where there would appear to be a sizable potential market? While the high cost of constructing cable systems in metropolitan centers is one deterrent, of greater importance are the regulatory actions of the Federal Communications Commission.

Influence of FCC regulation on CATV

An FCC freeze on VHF channel allocation was in effect from October 1948 to July 1952. One of its results was to encourage the early development of cable systems in communities lacking television service.[3] With the development of microwave relay systems, in 1953, cable operators saw an opportunity to expand revenues and profits by offering program diversity in densely populated metropolitan areas served by only one to two television stations.

This movement prompted quick opposition from the established broadcasting stations on the grounds that: (1) CATV competition tends to reduce the audiences of existing local stations, which, since advertising rates depend on the size of audience reached, would cut advertising revenues; and (2) CATV operators are not required to pay copyright fees on program material, which gives them an unfair advantage over local stations, who must pay for program material. The resulting controversy led the FCC, first, to assert control over the importation of distant signals by microwave relays (1965), and then to assume total jurisdiction over all CATV systems, including microwave operations (1966). Rules were issued prohibiting the importation of distant signals into the 100 markets with the largest television household population as established by a list compiled from a survey by the American Research Bureau (ARB). Cable operators also were not permitted to duplicate on the same day programs carried by local television stations.

In December 1968, these rules, which had discouraged the entry of CATV systems into the top 100 markets, were replaced by extremely restrictive regulations which put a freeze on new CATV installations and the importation of distant signals into the top 100 markets pending the adoption of a new set of rules. These restraints were criticized by the President's Task Force on Communication Policy.

As a result of the restrictions outlined above, there was little CATV growth in the top 100 markets. Cable operators claimed it was necessary to import outside programs in order to provide the diversity of entertainment which would attract subscribers already able to get a clear picture from existing local stations.

After considerable debate and controversy, the FCC issued a new set of rules, which went into effect on March 31, 1972, and were designed to permit CATV expansion and operation in major markets without jeopardizing over-the-air broadcasting.[4] Such rules provided that:

1. Systems in the top 50 markets may carry signals of at least three network and three independent stations.
2. Systems in markets ranked in size from 51–100 may carry at least three network and two independent stations.
3. All systems in the top 100 markets are entitled to carry two distant signals. (According to the FCC, permission to carry two distant signals not available in the community was given to enable CATV to attract investment capital and to open the way for the full development of cable's potential.)
4. Systems in markets below the top 100 in size are limited to three network signals and one independent.
5. Those outside any definable market are not limited in the number of signals they carry.

At the same time, over-the-air broadcasting was protected by the following rules:

1. Cable systems in the top 50 markets are prohibited from carrying any syndicated program for one year after its first appearance in any market and then for the life of the contract under which it is sold to a local station.
2. In markets 51–100 in size, different kinds of nonnetwork material would be protected for varying periods of time up to two years.
3. Same-time exclusivity protection was afforded network programming. (This, however, represents a reduction from a previous prohibition against same-day duplication.)

Potential liability for copyright fees

The controversy over whether cable operators should pay copyright royalties to program producers has not yet been resolved.[5] At issue is whether the carriage of a television program by a cable system constitutes a performance that makes it liable for royalty payment under U.S. copyright laws. In 1966 the U.S. District Court for the Southern District of New York ruled in the case of *United Artists Television, Inc.* v. *Fortnightly Corporation* (a CATV operator) that the cable firm had infringed on the United Artists' copyright when it retransmitted signals normally receivable in the subscribers' home viewing area. When the U.S. Supreme Court reviewed the case, however, it ruled in a five-to-one decision that under present statutes CATV incurs no liability for carrying copyrighted programs.

It should be noted, however, that the *United Artists* v. *Fortnightly* case involved a CATV system that carried only locally receivable signals. Accordingly, the Supreme Court decision on this case left unresolved the issue of whether copyrighted programs imported from distant stations by microwave constituted a "performance" and hence made the CATV operator liable for copyright payments.

That issue was addressed in a March 1973 Supreme Court decision in the *CBS* v. *TelePrompter* case. The Supreme Court held that cable systems do not have to pay copyright fees for carrying programs that originate at broadcast stations in distant cities.[6] The Court noted that "the signals that a CATV system receives and rechannels, have already been released to the public" over the air and therefore, the cable operator's importation of previously broadcast material does not constitute a "performance" under the copyright law.

The decision overturned a judgment by the Federal Appeals Court of New York which had distinguished between two kinds of activity by CATV systems, ruling that (1) when a CATV station retransmitted television signals normally available in the subscribers' home viewing area, this did *not* constitute a performance, but (2) when the same system retransmitted signals which could not normally be received on home sets, this *did* constitute a performance and accordingly made the cable operator liable for copyright payments.

The Supreme Court also commented that ultimately Congress must draft new copyright legislation to deal with cable television. The present Copyright Act was passed in 1909. Proposed legislation updating the Copyright Act has been held up in Congress since before the *Fortnightly* case was argued in the Supreme

Court in 1968. The TelePrompter decision encouraged a Senate Copyright Subcommittee, under the chairmanship of John L. McClellan, to move ahead with its task.

One of the controversial issues raised in revising the copyright legislation is whether Congress should establish the initial rates to be paid copyright holders to insure their "reasonableness." The new legislation proposes that cable systems pay the following royalties on copyrighted material:

1 percent of gross receipts up to $40,000
2 percent of gross receipts from $40,000 to $80,000
3 percent of gross receipts from $80,000 to $120,000
4 percent of gross receipts from $120,000 to $160,000
5 percent of gross receipts above $160,000 [7]

Operators of small, independent cable systems are especially anxious to be exempted from copyright fees, and they base their case on the grounds of hardship. As a means of gathering data on the question, Senator McClellan requested the Community Antenna Television Association (CATA), a national association of independent cable operators, to prepare a report showing the potential effect of the proposed copyright fees on their operations.[8] In compliance, CATA solicited financial statements from 1,000 operating systems not affiliated with the 25 leading Multiple System Operators (MSOs). From the 191 replies received, CATA divided the reporting systems into five categories based on size of subscriber totals. It then reported the collective revenues, expenses, and net revenues (prior to the repayment of debts and capitalization for system expansion) in each category. The results are summarized in Table 2.

Table 2

FINANCIAL SUMMARY OF
1,000 SMALL INDEPENDENT CABLE SYSTEMS, 1973

Number of Subscribers	Gross Receipts	Expenses	Net Profit	Net Profit per Subscriber
40 to 500	$ 298,470	$ 336,697	$ 38,227(loss)	$ 7.62(loss)
500 to 1,000	839,619	755,934	83,685	5.97
1,000 to 1,500	1,003,707	931,658	72,049	4.36
1,500 to 2,000	1,098,334	942,138	156,206	8.54
2,000 to 5,800	1,559,034	1,222,460	320,984	12.92

Source: *Broadcasting*, Dec. 31, 1973, p. 45, by permission.

CATA then computed by how much the proposed copyright fees would reduce the net revenues of each size category above. This was done by assuming the fee would be $.60 per subscriber, except in the 40–500 category, where it would total $.59. The computed reduction in net profit per subscriber is shown in Table 3.

On the basis of their calculations the CATA concluded that "if small, independent cable systems are not exempted from the monetary requirements of forthcoming copyright legislation, operations that are now 'just treading water' could be subject to disastrous financial setbacks."

Table 3

PROJECTED EFFECT OF COPYRIGHT LIABILITY
ON SMALL INDEPENDENT CABLE SYSTEMS

Number of Subscribers	Net Profit per Subscriber (without Copyright Payment)	Net Profit per Subscriber If Copyright Fee Is Paid	Percentage of Reduction in Net Profit
40 to 500	$ 7.62(loss)	$ 8.21(loss)	7.7
500 to 1,000	5.97	5.57	11.1
1,000 to 1,500	4.36	3.76	13.6
1,500 to 2,000	8.54	7.94	7.1
2,000 plus	12.92	12.32	4.7

Source: *Broadcasting*, Dec. 31, 1973, p. 45, by permission.

Note: Assumed fee: $.59 per subscriber for 40–500 category; $.60 per subscriber for all other categories.

Early in April 1974, the Senate Subcommittee approved the proposed new copyright bill and sent it to the parent Judiciary Committee[9] containing some provisions perceived as favorable to cable operators and some perceived as unfavorable. On the negative side was a provision prohibiting cable systems from importing live sports broadcasts from distant stations except when the same event is broadcast locally. Also, the fee scale that cable systems are to pay for the right to retransmit broadcast programs was regarded as too high. Cable interests had lobbied for a 50 percent reduction. Copyright owners, understandably, regarded the proposed fee schedule as too low. They had argued in favor of leaving the establishment of fees to arbitration and omitting them from the copyright law. The proposed legislation does provide for a review of the fees eighteen months after adoption of the bill. The National Cable Television Association had lobbied for the total exemption from copyright fee liability for cable systems with fewer than 3,500 subscribers, but this concept was not accepted by the subcommittee.

When the copyright revision bill passed through the Judiciary Committee on June 12, 1974, however, action favorable to the cable industry was taken on the objectionable features of the bill mentioned above. When the Committee "marked up" the bill, they completely eliminated the sports blackout provision and they also reduced the proposed copyright fee schedule by 50 percent.[10] The Committee did stipulate, however, that the fee schedule be reviewed by a copyright tribunal six months after the law is enacted and that it be reviewed at five-year intervals thereafter.

Although it was originally anticipated that the Judiciary Committee's report would be written about a week after the bill was marked up, and that the measure could get onto the Senate floor in July 1974, a request of the Senate Commerce Committee for time to review the bill delayed the movement of the legislation. The Commerce Committee, overseer of the FCC, which in turn regulates cable and over-the-air broadcasting, argued that copyright legislation should come under its purview.[11] The committee was, therefore, granted a 15-day referral period for consideration of S. 1361, the Omnibus Copyright Revision Bill.

Early in August 1974, accordingly, the Commerce Committee released its report on the proposed legislation. It recommended four amendments to the bill: (1) inclusion of cable systems in Hawaii and Puerto Rico in its compulsory licensing scheme, (2) directions to the FCC to promulgate sports carriage rules for cable systems, (3) exemption from copyright liability for pre-1972 systems which served communities that otherwise would not receive television service, and (4) exemption of broadcasters from performance royalties for recorded music. Finally, the Commerce Committee stressed that its actions on S. 1361 should not be interpreted as an endorsement of that bill in any form—its original form, as it had been amended by the Judiciary Committee, or even with the amendments recommended by the Commerce Committee itself.

Because the Commerce Committee did not take action on the copyright revision bill until August 1974, and because congressional involvement in the Nixon impeachment activities was then delaying legislative activity, observers believed the likelihood of House action on the proposed legislation during the Ninety-Third Congress was nil. As of this writing, therefore, the copyright question remains unresolved. Cable operators who are considering entering one of the top 100 television markets or expanding their facilities, if already in operation there, still face an important uncertainty as to their potential liability for payment of copyright fees. How soon this uncertainty will be resolved is difficult to predict, but in the meantime its continuation tends to have a negative effect on plans to increase CATV penetration in the top 100 television markets, and it is believed to impair the ability of the cable industry to attract the capital needed for substantial growth.

Local regulation

Earlier discussion indicates the influence of FCC regulation on CATV development and operations. Cable operators hoping to enter, or already serving, the top 100 television markets are also subject to franchising and regulation from city governments. In order to build a CATV system in a major market the cable operator must be awarded a franchise by the local government, which is likely to carry with it the payment of a stipulated franchise fee to operate in the community. Local authority to franchise and regulate cable television, which extends to services offered and rates charged as well as other aspects of the system's operation, derives from the cable system's need for access to city streets and other rights-of-way, but it must be exercised within the framework of federal and state laws and regulations.[12]

Municipal franchise authorities must follow certain standards if their franchises are to obtain an FCC Certificate of Compliance—without which a cable system cannot carry any broadcast signals. In granting franchises the authorities must consider legal, personal, financial, technical, and other qualifications of applicants by means of a full public hearing. The FCC rules also require that the franchising authority approve the rates charged to subscribers—i.e., the rates established initially and, as well, subsequent revisions to rates in response to changing cost or demand considerations. This provision limits the freedom of action of the cable operator in establishing rates that will foster penetration of the market and still provide an adequate return on his investment.

Significantly, under the FCC rules the franchise may not prohibit pay television. It must make provisions for handling subscriber complaints. The initial franchise duration may not exceed fifteen years. Also the cable system must begin construction within a year after the FCC has issued its certificate of compliance. It must wire a "substantial percentage of its franchise area each year" (20 percent is suggested). Finally, the franchise fees paid to the municipality cannot exceed 3 percent of subscriber revenues without specific approval of the FCC.

State regulation

With the growth of CATV, several states have also taken steps to regulate certain aspects of cable operation. According to Morton Aaronson of the Massachusetts Cable Commission, in 1973 there were three states that had separate state commissions charged with regulating cable. Massachusetts was the first, New York was the second, and Minnesota had then recently enacted legislation setting up a separate commission.[13] In 1973 approximately seven other states had Public Utility Commissions with cable bureaus, and legislation with regard to state regulation of cable television was pending in various degrees in about 30 more states.

According to Aaronson, state commissions make contributions which are beneficial to both the public interest and the interest of the cable industry. In Massachusetts, for example, the commission found governing city officials lacked understanding of cable television. Here the function of the state commission is to provide information to guide local authorities in carrying out their franchising and regulatory activities. Such guidance, he claims, helps to assure that cities and towns make adequate provisions regarding public access to the cable system and that local origination of programs is encouraged, as well as furthering the discharge of other civic responsibilities.

In terms of the cable industry, the guidance of the state commissions can serve to make sure cities and towns are not unreasonable and unrealistic in what they require from the cable operator. A state commission can also help the cable industry gain access to pole rights. In both New York and Massachusetts, for example, the cable commissions were taking specific steps in 1973 to see if they could evolve regulations which would give the cable industry reasonable access to pole rights. Most state utility statutes require that the pole rental fees which telephone and power companies charge cable systems for attachment of their cables be regulated and held at a modest level. This is especially important in some areas where utilities, in efforts to restrict CATV operations, have raised pole rentals by several hundred percent.

The trend toward state regulation has important implications for the cable operator who is in, or planning to enter, one of the top 100 television markets. Of special significance in state regulation is the question of whether CATV systems are to be placed in the same class as any telephone or power utility and subject to the same rate-making procedures they are.

In discussing the question of state regulation of CATV systems, Archie Smith, chairman of the Rhode Island Public Utilities Commission, made some significant comments at the 1973 National

Cable Television Association (NCTA) Convention. As Public Utilities Administrator, Smith has the task of awarding all cable franchises in Rhode Island and regulating CATV operators. He argues that "regulation of the CATV industry is inescapable if we are to assure proper and safe construction of plants and maintenance of systems and uninterrupted viewing of wholesome and worthwhile programs at reasonable charges."

He then adds:

> For regulation to be effective and for the future CATV requirements of the public to be met, cable regulation must reward excellence in management.... This calls for incentive regulation, not the computation of original cost or fair value and the application of a fixed rate of return to such figure. That is the usual type of rate regulation. Neither is the common-carrier type of regulation based on operating ratios appropriate. Cable communications, being in its infancy, must be encouraged to experiment and innovate in all areas of its operations. Involved regulatory rate hearings with excessive procedural requirements would tend to inhibit new services. Up to this time the market place has adequately regulated cable rates. When rate regulation becomes necessary, rate control will be enacted in the same manner that rate control of electric, gas, and other utilities occurred, that is, when there was a very high degree of market penetration.
>
> For the present, rate regulation should restrict itself to an oversight or surveillance function to assure the public that rates, both for subscription and leasing of channels, are published and subject to review to assure that they are not discriminatory and are uniform for various classifications of users and uses.[14]

Also important to the cable operator is the issue of whether the state regulatory body imposes a tax on his operations. The Connecticut Public Utilities Commission, for example, proposes an annual 6 percent utility tax on cable system gross income.[15] This tax appears sizable when juxtaposed with the FCC rule that franchise fees paid to the municipality cannot exceed 3 percent of subscriber revenues without specific approval of FCC.

In summary, as we discuss elements of marketing strategy appropriate for building penetration in the top 100 television markets, we must keep in mind that the cable operator interested in doing business in these markets is subject to regulation by the FCC and also by a state regulatory body or the local municipality in which his system is located or by both, and his decisions about building and operating must be made within the restrictions promulgated by these agencies.

Special problems in large urban markets

Cable operators face a difficult challenge in developing a profitable business in the large urban markets. While Manhattan is a special case, discussion of problems encountered by Tele-Prompter Cable TV and Sterling Manhattan Cable Television in developing this market may be useful to operators who receive franchises in Chicago, Detroit, and similar cities lacking CATV as they develop plans for opening these markets.[16]

Off-air reception of television signals is mixed, at best, all over Manhattan, and this situation provides the strongest single inducement for television viewers to subscribe for cable service, since, with over-the-air broadcasting already providing seven VHF signals and five UHF stations, consumers have little to gain in program diversity in New York City.

Sterling Manhattan's experience shows that only about 30 percent of potential subscribers (those living in buildings into which cable has gained entry) will sign for the cable service. As long as better reception remains the principal attraction, this proportion will not rise substantially. By October 1973 Sterling Manhattan had signed up 59,000 of the 185,000 potential subscribers passed by its cables—a 32 percent penetration. TelePrompter Cable Television had gained 55,000 subscribers out of 253,000 passed, a penetration of 28 percent.

These systems face special problems that tend to run up their costs in providing service. One is the city franchise requirement that cable be placed underground. The main trunk lines of the cable companies share already existing under-the-street conduits with other utilities. A second important problem, which has proved to be the most expensive of all, is the New York City landlord. In Manhattan, right of entry must be gained to an apartment building before individual households can be given service. In addition, entry to a block must be obtained before the various apartment buildings can be wired. A few recalcitrant landlords can keep a whole block from getting service by refusing entry from the trunk line in cases where there are no reasonable alternate entry points.

There is also a problem with absentee landlords, particularly in Harlem. They must be located, often at considerable expense, before the cable can get to its customers. Such obstacles slow down and limit penetration.

Then too, some landlords use their gatekeeping power to secure a share of the cable television revenue. Landlords of buildings which the cable operator is especially anxious to enter have successfully held out for as much as 5 percent of monthly subscription fees paid by the building tenants.

An additional major marketing problem in upper Manhattan is the fact that a third of the people move at least once every two years, a rate substantially above the national average. In the average franchise outside New York the subscriber retention factor is high. The cable service in such areas is almost like a telephone, people keep it for a long period. In upper Manhattan, by contrast, as the moving vans arrive and depart the selling job has to be done over and over.

While it is true that developing the top 100 markets poses challenging problems of cost, financing, and marketing, such markets also offer the greatest potential to cable operators who approach them wisely. A Samson report, for example, highlights the following opportunities:

> Residents of large metropolitan centers live closer together and generally spend more on entertainment than their suburban and rural counterparts. Higher population densities and higher incomes and educational levels mean more potential subscribers per mile of the distribution cable, thus lowering the cost-per-subscriber for the cable operating system. Poor TV reception in the city plus the desire for additional programs as an information and entertainment source offers an attractive market for the system with extra channel capacity.[17]

Outlook for Future Growth of CATV

Now that FCC regulations which are designed to get cable moving without jeopardizing over-the-air broadcasting have been promulgated, what is the future outlook for CATV? In 1974 there were 8.1 million cable television subscribers—12.5 percent of U.S. television households. In its 1971 projections of future growth of cable television, the Sloan Commission estimated that the number of cable subscribers in 1980 would probably range between 29 million and 37 million, depending on the number of viewing alternatives available by the end of the period (see Table 4). This would translate into household penetration figures of from 48 percent to 61 percent of U.S. television households.[18] On the basis of this analysis the Sloan Commission projected household penetration by 1980 of from 40 percent to 60 percent.

The Sloan Commission projections were made at a time of considerable overoptimism as to the future of CATV (1971) but before the FCC promulgated the new rules in March 1972 which were designed to relax the freeze on cable television. In a 1973 study for the National Science Foundation, Walter S. Baer analyzed the growth of CATV households in the United States over the previous decade.[19] He found that the growth in cable subscriptions had averaged 23 percent annually during that period. Projecting this rate of growth to 1980 he arrived at an estimate of 30 million cable subscribers, or 42 percent of U.S. television households for that year. This would be at the bottom of the range

Table 4

ESTIMATES OF ANTICIPATED GROWTH IN CATV,
1974 TO 1980

Item	1974	Est. 1980	
		Sloan Commission	Baer
Number of subscribers (millions)	8.1	29 to 37	30
Household penetration (in percentage)	12.5	48 to 61	42

Sources: "Short Course in Cable, 1974," *Broadcasting*, April 22, 1974, p. 23 (1974 data); Sloan Commission, *On the Cable* (New York: McGraw-Hill Book Co., 1971), p. 215; Walter S. Baer, *Cable Television: Handbook for Decision-Making* (Santa Monica, Calif.: Rand Corporation, Feb. 1973), pp. 7–10.

estimated by the Sloan Commission (40 percent to 60 percent). The National Cable Television Association, an industry group, has predicted a lower range of 35 percent to 40 percent.

Baer notes, however, that cable growth during the period 1971–72 was slower than before. The 1971–72 percentage growth when projected to 1980 results in only 15 million cable subscribers, or 21 percent of television households. According to Baer, therefore, estimates of cable subscribers in 1980 range from 15 million to perhaps 44 million with the most likely figure being 30 million, or a penetration of 42 percent of television households.

In a paper published in February 1974, Baer commented further on trends in cable subscriptions.[20] He noted that in less than two years the American mood had changed sharply from unreflective optimism to skepticism or downright pessimism about the future of cable television. He cited industry overexpansion, high interest rates, and projections of a general economic downturn as factors which had led cable companies to cut back their plans for construction in the major U.S. cities. He reports that some companies had even backed away from accepting local franchises which they had spent many thousands of dollars to win.

According to Baer,

...It is likely that the pendulum has once again swung too far. Cable television still represents a growing force in the American communications system—a force that could bring significant changes in society's use of communications in the next two decades. The Communications

Revolution promised by cable has not been thwarted, but rather slowed to a more evolutionary pace.

Continuing, he notes that the 23 percent annual growth rate in cable subscriptions during the decade prior to 1971 was beginning to level off.

Fewer new cable systems began operating in 1972 than in 1971, with a similar downward trend likely for 1973 when the statistics are compiled. Most systems in areas of poor broadcast TV reception are now well saturated; usually over 50 percent of the homes with access to the cable already subscribe. Consequently, there is little room left to grow in these areas, and further growth must be in the major population centers, where construction costs are high, competition with broadcast television keen, and government regulation more restrictive. Cable has not yet effectively penetrated these major markets, which contain more than 80 percent of the nation's population. As a result, many past projections of cable subscription growth in the 1980s.... seem overoptimistic. The Sloan Commission predicted that 40 to 60 percent of the nation's households would be cable subscribers by 1980. Today, the industry would be happy to achieve half that total, but even a 25 percent estimate for cable penetration in 1980 may be too high.

Another viewpoint is expressed by the U.S. Department of Commerce in its publication, *U.S. Industrial Outlook, 1974,* released in late October 1973.[21] In its review of the cable industry, it predicts that "subscribers will continue to increase at an annual rate of 16 percent, numbering 23.5 million by 1980.... The outlook for cable television is bright provided that the very substantial capital required is available on reasonable terms and that undue delays are not encountered in the issuance of franchises and FCC certificates of compliance."

The projections reviewed above were formed on the basis of certain assumptions as to the comparative service offered by CATV versus over-the-air broadcasters, growth in the proportion of families with color television receivers, anticipated changes in average household income, size of installation and subscription fees charged by cable systems, and the offering of specialized consumer services by CATV without competition from traditional broadcasters, among other considerations. In the pages that follow, certain of these influences will be examined in depth, together with other important factors which will tend to influence the long-run performance of CATV in penetrating television households. This analysis will identify the problems that must be solved by cable television operators if the industry is to achieve its full potential, and it will also identify the opportunities that may be grasped by wise decision making, sound planning, and effective execution of appropriate strategies.

Franchise grants and certificates of compliance

Before new or expanded cable facilities may be put into operation in the top 100 markets not served by CATV, municipal or
state authorities must issue a franchise to cable operators and then
the FCC must issue a Certificate of Compliance. When the FCC
relaxed the freeze on cable entry into center-city areas in major
metropolitan markets in 1972 only a few such areas had cable
systems. These included New York City, Los Angeles, San Diego,
San Francisco, and Seattle. In almost every case, however, there
were suburban cable systems surrounding these major cities, and
their operators, as well as many other aspiring firms, had submitted applications to franchising authorities for center-city systems.
Franchising of these systems was expected to proceed slowly
because of the intense rivalry between competing applicants for
the right to serve these markets. All of the 100 major markets are
likely to be served by cable systems. Projections by the Samson
Science Corporation about the growth in total U.S. cable television systems from 1972 through 1980 are shown in Table 5.

Table 5

PROJECTED TOTAL U.S. CABLE TELEVISION SYSTEMS,
1972–1980

Year (As of January 1)	Net Increase	Total Operating Systems
1972	145	2,760
1973	150	2,910
1974	180	3,090
1975	210	3,300
1976	185	3,485
1977	170	3,655
1978	155	3,810
1979	140	3,950
1980	120	4,070

Source: *Cable Television: Takeoff into Sustained Growth* (New York:
Samson Science Corp., 1972), p. 45.

In May 1973 Don Andersson, Director of Statistical Services,
NCTA, reported that the FCC had issued certificates of compliance to 119 proposed new systems in the top 100 markets,[22] all of
which under FCC rules are required to complete significant construction within one year from the date of the certificate. As of
May 1973 certificates had been issued for the following major
cities:

Philadelphia, Pennsylvania—the Number 4 market
St. Louis, Missouri—Number 11
Columbus, Ohio—Number 27
Tulsa, Oklahoma—Number 24
Chattanooga, Tennessee—Number 78
Jackson, Mississippi—Number 77
Memphis, Tennessee—Number 26
Moline, Illinois/Davenport, Iowa—Number 60
Sioux Falls, South Dakota—Number 85

According to Andersson, the total population in these grant areas was approximately 7.8 million. At 3.1 persons per home this translates into more than 2.5 million housing units. Since 13 million homes were estimated to be passed by cable lines as of 1973, the construction of systems in front of 2.5 million additional homes would mean an increase of 19 percent in the potential market (2.5 million divided by 13 million).

Time required for development

Cable penetration of major urban markets takes time,[23] not only because it depends on the action of regulatory bodies in franchising the operator and issuing a certificate of compliance, but also it takes time to construct the cable system and prepare it to serve the viewing public. Preparing an initial proposal to compete for a franchise may take from one to six months. Municipal action on the franchise may require from one month to more than five years. Construction time depends on the difficulty of construction, the cooperativeness of existing utilities, the availability of funds, and subscriber acceptance of the new system. Available capital may dictate that the system be constructed in segments, with cable service offered to home owners passed by the first segment before construction is begun on the next segment. Success in getting consumers to subscribe to cable service may furnish a portion of the funds to finance further extension of the system.

According to industry observers, once a system is installed and the initial subscribers are connected, profitable operation is approximately three years ahead. An effective marketing strategy which will build subscriber penetration rapidly past the break-even point is clearly of greatest importance.

Services Which Build Penetration

Sound marketing strategy requires that management offer the consumer services which will give the cable system a competitive edge. Let us examine the competitive situation in the major urban markets, then consider what combination of services is most likely to lead television viewers to subscribe to cable television.

The generally good reception of over-the-air broadcasting in most large urban markets negates one of the key benefits that initially led to the establishment of community antenna systems, although major cities where skyscrapers often completely block or distort television signals, such as New York, are exceptions. The increasing proportion of television viewers owning color sets, however, has made users more particular about the quality of reception. The new owner of a color set often faces the necessity of installing a more costly antenna than that which sufficed with a black-and-white set. If cable television service is available, he may choose this alternative as a means of getting the desired quality of reception. Color-equipped homes in the United States as of November 1972 totalled to 38.3 million, or 59 percent of all U.S. television homes.[24] Samson Science Corporation estimates that by 1982 60 million homes or 80 percent of all U.S. television households will have color sets.[25] This trend should tend to stimulate CATV penetration of major markets.

Another benefit that led television viewers in medium-sized and smaller communities to subscribe to CATV was the availability of additional programs through cable service. In many of the top 100 markets television viewers already receive all three networks, one, two, or three independent stations, plus an educational outlet. Where this is true, cable has little to offer in additional program alternatives. Clearly the character of over-the-air television competition in each of the major markets is an important consideration in estimating the likelihood of achieving a profitable market penetration.

In this connection, the analysis of Greg Liptak, LVO Cable, Tulsa, Oklahoma, is noteworthy.

I think that selling cable TV in the nation's top 100 markets will be a difficult job. I think it will require a cable system which delivers pictures of excellent quality, combined with as many services as is possible to deliver.

I'd like to confine my comments to a competitive marketing situation found in 55 of the nation's top 100 markets, where all three television networks, as well as an educational station, are available off-the-air in good quality, probably on "rabbit ears."..... In this type of market, no independent television service exists.

The key question, I think, at the present state of our industry, is: How well will cable do in the competitive environment, i.e., full network service and an ETV?...

First of all, I don't think it's possible to operate a really successful cable television system in this competitive environment with just two distant independent signals and a couple of automated services.... I don't believe it's possible to achieve an operating level of success, and defining that as 40 percent penetration of the market, with this type of limited service package. I think in this competitive market situation, every practical service must be presented.

I think we have a major factor going in our favor: I think people in America today are fed up with the number of commercial interruptions on commercial television.... Further, I think people are disgusted with the early start of reruns on the networks.... Because of these factors, cable has a tremendous opportunity today to provide an alternate choice for the people of our nation.

Obviously, distant independent stations, ... will be particularly valuable where no independent service is currently available off-the-air. If a cable system is fortunate enough to have good independents authorized—and ... the quality of programming varies widely between the independent[s]—then that cable system is off to a running start in providing a new alternative to its customers. If the independents that are authorized for given cable systems do their job well, or by counter programming, then these signals will be, in my opinion, the principal sellers on our new cable systems in the nation's top 100 markets, particularly in the 55 markets where no independent service is currently available....

Other selected services such as the weather channel, news, stocks, and so on, fill a definite need in our cable systems, but taken individually, they're not key factors in getting consumers to buy cable. An aggregate, ... is important.

Obviously, other local programs, particularly sports, will be significant; other specialized channels such as the religious programming developed and currently in use by a number of cable systems ... have [some] appeal.

We must, I believe, strive to develop as many of these specialized subscriber services as we possibly can, and we must see to it that these services are programmed at convenient times for our customers. We must not overlook the attractiveness of pay cable services in getting subscribers into our major market systems.

My conclusion is that cable systems can achieve success, defining success as initially attracting subscribers near the 40 percent penetration level in competitive environments where full network service and an educational station are available off-the-air.... assuming, of course, an excellent service honestly rendered and at the right price.[26]

While bringing in attractive independent signals may be an important benefit to offer viewers in the 55 markets discussed above, what marketing strategy must be followed to attract subscribers in the remaining 45 major markets where three network stations are available, plus one, two, or three independent stations, as well as one educational signal—all received with rabbit ears?

Bill Pitney, Cox Cable Communications, Atlanta, Georgia, illustrates the challenge of such a competitive situation in discussing the problem his firm faced in 1973 in St. Louis, Missouri.

> ...We've been certified for two distant independent signals, one from Kansas City, and one from Bloomington, Indiana. St. Louis has three networks and two fine independents, and I can't conceive that an independent from Kansas City or an independent from Bloomington is going to be any significant factor in that market as far as attracting a viewing audience. So we had to take a different approach and do some different thinking.
>
> We don't have the answers yet, but we are going to start some construction this year. And I'm very hopeful that by the time that I have to turn that plant on, that I have found some premium TV that will generate some revenue for me.
>
> . . . .
>
> I don't think anyone knows what premium TV will do for you. It needs to be tested yet. There are some tests going on but they're not conclusive at this point in time. If I can buy enough time before I have to open the St. Louis market, maybe we will have found an answer to this.
>
> I think that it's a big part of the future of cable TV. We've got to have it. We've got to have something to fill in the gaps between now and ten channels off a satellite. Again I don't have the answers but we are certainly looking for them.[27]

After the panel discussion during which Liptak and Pitney made the points cited above, Paul Kagan, of Paul Kagan Associates, Incorporated, New York City, underscored the problem in the following comments:

> In the suburbs of New York, for example in New Jersey, where Columbia Cable has systems operating, this is not a distant signal equation at all. ...What do I do when I get into a top 100 market where distant signal importation is not going to be in the picture? In Columbia Cable's case, they've reached saturation levels of anywhere from 30 to 40 percent based on conventional cable, you know just what's in the air, and the Knickerbocker-Ranger package.
>
> I know of other systems in metropolitan areas that don't even have a Knickerbocker-Ranger package that have been able to achieve a 30 percent saturation level based on the fact that some people will take service because of some reception problem that they have in their locality. Some people will take it on a status basis. Some people will take it on an automatic channel basis, as I would take it just for the stock market ticker. Because $5 per month for a stock market ticker is competitively priced with any other stock market service available.
>
> So if we address the subject perhaps not at the level of what do we have to do to achieve 40 percent saturation, but on a situation where we can reach a reasonable cash flow level, such as 30 percent saturation in the suburbs of a major metropolitan area....

If we are in a situation like that, and we can reach 30 percent without distant signal importation, then comes the fun: How do we get to make money in this system? What can we market in this top 100 market, meaning no distant signals?[28]

In response to this question, Bill Pitney said,

...I think that we have to turn our attention to premium TV in those markets. And I don't mean it has to be a supplement to that market. We've got to find some premium TV to go into that system, because you cannot depend on distant signals' making the market for you. ...I think for present-day operation in some of the top 100 markets, the distant independent signals will sustain you until such time as pay TV of some form develops a little farther than it has until now.

In view of the importance attached to importing signals of distant independent stations in 55 of the top 100 markets, during the foregoing panel discussion Don Andersson of NCTA called attention to certain "economic and regulatory facts of life" which are unique to the major urban markets. His comments relate strictly to the "non-grandfathered" proposed new systems in the top 100 markets, subject to the restrictive regulations spelled out in the FCC's cable television regulations of March 1972. According to Andersson,

There are two aspects of these rules that have significant bearing on the development and marketing of big city television systems.

First, the signal carriage rules and some subsidiary matters relating to them. Second, those regulations which impose cost burdens above conventional system costs. The carriage rules are deceptively simple. You can carry the in-market signals and those which are significantly viewed, and, with an exception of a handful of markets within the top 50 with three independent stations that can be imported, you can reach out to import two independents. If you can stay away from those in the top 25 markets, you can go anywhere to bring in your independents.

But, if you choose, as nearly all the operators are doing, the major independents in the top 25, you are restricted to a choice of one or both of the two nearest top 25 markets. So, the benefits received under the carriage rules are thus restricted by a mileage factor. They are further eroded under the exclusivity rules which require varying degrees of protection for syndicated programs in the home market, differing in the length of the protection period, depending on whether your markets are 1 to 50 or 51 to 100.

Several attempts have been made to ascertain the severity of imported program losses that exclusivity protection will influence. The Rand Corporation came up with a study revealing that in the larger markets, those with two or three independent stations, imported signals would be unavailable 50 to 65 percent of the time. Now, this applies to about 18 of the 50 markets.

For all other markets in the first 50, signal[s] would be unavailable 20 to 40 percent of the time. In markets 51 to 100 there would be no signal for about 16 percent of the time.

Now, on the same day that they issued the new rules, the FCC also offered up a new proposed rule-making, which, if effectuated, would prohibit the importation of a televised sporting event into a market if a home [town] sports team was engaged in a similar sporting event on the same day. This could have a serious effect on cable TV in many of the major markets.

As for the increased costs, the rules dictate there is a requirement for a twenty-channel system ... that it be a two-way facility. And there are the costs of access channels. For most markets there will also be the cost of microwaving in those distant signals. In some markets, if the cable system chooses to select from the two nearest top 25, where the best independents are, it could mean reaching out for signals in markets that are more than 900 miles away. Unrelated to the rules are the increased costs per mile of plant anticipated in major city construction through the necessity of undergrounding in many of the middle city areas.

So, on the one hand, we have a system that will cost more to build; on the other hand, we won't be able to offer much more station programming than is already available in the top markets. CATV's historic pitch, "There is more to see on cable TV" has been considerably muted by the FCC.

Certain markets, however, because of their proximity to others will be able to draw on the more profitable independent stations in key markets which carry the home and away games of several major league sports teams; which have the where-with-all to buy the better film packages, and which can afford the higher caliber syndicated shows. Other markets will have access to more inferior programmed independent stations.[29]

Underscoring the above comments is the analysis presented by W. Bowman Cutter, Executive Director of the Cable Television Information Center, at the NCTA convention in April 1974.[30] Cutter stated that, under existing FCC regulations, cable is being denied an attractive commodity to market. "Entertainment of some sort is the basic vehicle of cable growth," he noted, acknowledging that at present, however, "cable just doesn't have a service to offer." He claimed that with the present limitations on the number of television signals a system can bring into its market, the extent of cable penetration might not surpass 25 percent of the nation's television households. But if just four more signals were permitted each operator, Mr. Cutter speculated, that expectation would double.

One of the principal problems of today's regulatory environment, according to the Rand Corporation's Henry Geller, is that, while the FCC's present cable rules were designed to encourage

cable's growth in the top 100 markets, economic limitations coupled with the problems cablemen have experienced in marketing
the services made available to them by the commission have
resulted in a virtual moratorium on viable development.

The challenge that confronts cable operators in developing
services which will appeal to viewers is indicated by Alfred Stern,
Chairman of Warner Cable Corporation, New York City, in a talk
delivered in December 1973.[31] Mr. Stern said venturers had
"learned that cable in the larger markets is not an essential
commodity." In areas that already enjoy a multitude of communications and leisure-time services, cable is "more an optional
added luxury than an electronic necessity."

The industry has also discovered that major-market subscribers,
unlike their small-town counterparts, will not cling to the medium
if the service proves disappointing. This, he acknowledged, has
caused an "immense turnover problem."

Beyond that, said Mr. Stern, it "could be quite a while" before
specialized cable services "are ready for delivery on a large scale"
in the big markets. "The hardware isn't there and the software
isn't there," he said. "The hard truth is that the development of
new services is still merely in the experimental stage."

In spite of this pessimistic view, certain cable operators have
been successful in selected urban markets. At the Chicago NCTA
convention in April 1974, for example, Ed Drake of LVO Cable,
Tulsa, Oklahoma, reported on his firm's recent successful entry
into Tulsa.[32] In this instance, the cable system was introducing
that market's first independent service-stations from Dallas in the
spring of 1974 and Kansas City later. Initial saturation was reported at 58 percent.

While the LVO system was being built at an urban cost (eventually over $15 million), the Tulsa experience was not regarded as
valid for a Boston, Oakland, or similar city by Paul Kagan, according to comments of his reported in *Cablecast*. Kagan did point out,
however, that "a considerable number of Tulsas" were "yet to be
built,...one big reason why the cable industry gave off an aura of
confidence amid their troubles in Chicago this year [1974]."

He noted that

> American TV & Communications, for example, is providing initial
> independent service in Columbus, Ohio, where it expects to have a
> huge financial success (12-channel, single cable, no converters), and
> Albany, New York, where it plans to turn on its system May 2, 1974.
> . . . .
> It is also netting 40 percent saturation on first passes through Or-
> lando, Florida, where it has turned the debut of the market's first UHF

independent into a plus by promising immediate reception of the temperamental UHF signal. ATC is also planning major new construction in Fresno, California, Durham, North Carolina, and San Diego, California, and seeks franchises in Spokane, Washington, Shreveport, Louisiana, and Roanoke, Virginia. (The list is really longer. ATC has no dearth of projects and appears capable of financing all of them.)[33]

Kagan cites ATC as an example of a firm that has been successful in spite of recent difficulties in the industry,[34] making sound acquisitions over the years, conservatively avoiding over-extending itself in the wrong markets, and generally hitting just about all of its internal targets. ATC's earnings are up, and since its inception in 1968 it has sustained its annual subscriber growth at 20 percent.

Kagan lists as keys to ATC success "its very careful picking of markets in which to operate, and the way it has handled its finances." Rather than falling in love with the biggest of cities following institution of the FCC's 1972 cable rules, President Monty Rifkin is quoted as saying ATC "adopted a more conservative approach and looked to those potentially viable markets where the addition of two or three independent signals could make a meaningful impact on viewer choice."

According to Rifkin

There are approximately 14.5 million television homes located in markets which do not receive any independent station signals (approximately TV markets 30 through 100) with only 1.6 million of these homes currently served by cable....

Significantly, of nineteen franchises ATC is operating or developing, seventeen fall within markets 27 to 100. More than 800,000 homes in presently franchised areas are not yet passed by cable and more than 500,000 of these receive no independent TV service off-the-air. It all translates into years of future internal growth.[35]

According to Kagan,

Despite its early leveraged look, and the big construction that lies ahead, ATC still has $15 million of unused credit lines, is not in need of new equity or debt financing, and could conceivably construct all present franchises without a return trip to the money markets.

When it went public five years ago, ATC had 100,000 subscribers in forty-four communities in seventeen states. Today it serves nearly 430,000 customers in 130 communities in thirty-one states.[36]

Consumer reactions to CATV

In determining what CATV services are to be offered in a major urban market, a key consideration is the demographic characteristics of the television viewers to be served. Interesting data were gathered in one CATV market by Louis E. Boone in 1969. They are summarized below:

> The objectives of this research were to determine whether the Consumer Innovator and Consumer Followers of Community Antenna Television Service could be identified on the basis of distinguishable socioeconomic characteristics and personality traits and to identify these characteristics possessed by Consumer Innovators which distinguish them from later adopters.
>
> The Consumer Innovator possesses different personality traits than does the later adopter of CATV service. He tends to exhibit more leadership potential, be more socially mobile, possess more self-confidence, a greater acceptance of newness, and higher achievement levels than the later adopters.
>
> The proving of these hypotheses points to the existence of "Consumer Innovators" and the possibility of identifying these first buyers in any community. This should allow the market planner to segment his market and utilize this group as a test of his product's acceptance. Rather than using whole cities as test markets, it may be possible to focus on Consumer Innovators and observe their purchase behavior. Also it should be possible to utilize the information regarding their personality traits to construct promotional campaigns designed specifically to appeal to these individuals.[37]

One of the benefits CATV service offers consumers is a greater diversity of programs than is available over-the-air. Accordingly, it is significant to examine whether CATV viewers actually utilize the increased number of programs available to them. This question is examined by Melvin A. Harris in a study published in 1971.

> This study investigates television consumption behavior in terms of channel use as related to channel availability. The study examines whether television consumers make or will make use of all available channels in a multiple channel television system....
>
> Four hundred eighty households with all channel reception capabilities were randomly selected from the metropolitan areas of eight different markets surveyed by the American Research Bureau during the fall of 1970.... Respondents chose programs from projected program schedules to indicate channel use.

The findings of the study were:

1. Across all current channel availabilities, consumers use only three to six channels.
2. When projecting program viewing from a six, twelve, or eighteen channel prime time program schedule, consumers use only about four channels.
3. High consumers of television use more channels than low consumers.

4. Channel use increases as the number of available channels increases, but not in proportion to the increase of available channels.
5. Commercial network services are the most used channels, with additional channels being used more by high consumers.
6. High users of a program guide project use of more channels than do low users of a program guide.

Conclusions reached are:

1. The average television consumer does not use all the television channels that are available to him, as most of his use is limited to the three network channels and one or two public or independent channels.
2. The average television consumer would not use many more channels, even if twenty or forty channels were available.
3. An increase in the number of available channels would lead to a fractionalizing of the audience, mostly among the non-network channels.
4. In terms of audience size, the importation of additional channels into a market by cable would be most detrimental to independent and public channels.
5. The segment of the audience that uses public and independent channels the most is the segment that watches television more than average.
6. In terms of overall television use, public television stations are not primarily serving a neglected audience, as the majority of consumers who used public television channels watched televison more than five hours per day.
7. Independent television stations should promote the use of their channel rather than just the viewing of any particular single program, because specific programs may well be ignored if the channel is not regularly used.
8. Sales practices of independent television stations should emphasize the unique characteristics of their audience in terms of amount of television viewing, age of household head, and family size.[38]

Importance of good programming

While considerable emphasis has been placed on the ability of CATV operators to import signals from distant independent stations with high quality program schedules, it is not enough for CATV operators to rely on retransmission of programs of over-the-air television stations in their efforts to build subscriber penetration. They should also give high priority to the origination of programs to be cablecast over their own channels. Dr. William Melody of the University of Pennsylvania made the following comments on this point before a state regulatory convocation in June 1974:

> Communications opportunities opened by cable will not be developed by the simple creation of capacity. They will depend directly on the resources committed to software. In fact, despite the substantial investment required to wire the nation, we must anticipate that full development of cable opportunities will require an even greater investment in software.[39]

The significance of good, locally originated programming is suggested in the assessment of cable's 1974 prospects appearing in the *Broadcasting Cable Sourcebook,* 1974. This article notes that during 1973 "the industry began to look around for ways to augment the monthly subscriber revenues and increase profitability. First on almost everyone's list of possibilities was pay cable. The technology for delivering movies and sports for pay bloomed during the year, and cable operators began to have visions of added revenues, increased saturation levels, and a means to crack the signal-saturated urban markets."[40]

However, in October 1973 Edwin A. Deagle, Director of Analysis at the Cable Television Information Center, Washington, D.C., questioned whether subscription programming and satellite interconnection will save the cable industry. The future for cable Dr. Deagle contends, is not in entertainment but in the medium's potential as a substantial vehicle for broadband distribution systems. He believes pay cable cannot sustain the industry, because "pay cable is another form of entertainment. What people want in television is mainly what they're getting now"—i.e., the general-interest fare currently available on the networks.

Cultural and theatrical presentations lack sufficient mass appeal; major sports events are already on television; and films, he asserts, "may draw a few more people into the industry but not a hell of a lot more."

The main problem, Deagle feels, is that the number of television households—now placed at around 65 million—has nearly reached a peak. Thus, if the number of distribution sources is increased by pay cable and satellites, the audience for all programs will decrease.

...A "much stronger definition of market analysis must be produced," he says. "And that analysis will show cable's marketplace falling within two major categories—closed-circuit distribution to institutional users and home terminals...." Subsequently, he forsees a "tremendous amount of investment" in cable systems as entrepreneurs become aware of their potential in digital communications. That investment should result in a few new systems employing a full range of the advanced technologies by 1980. And from there, Deagle says, the "neighbor effect" will take over. As consumers become aware of the advantages those technologies offer, cable "will catch on very rapidly."[41]

Conclusion and Report Outline

The outlook for future growth in CATV, then, depends on the success experienced by CATV operators in increasing subscriber penetration in the top 100 television markets. It is clear that the CATV operator must be consumer oriented in planning a marketing strategy to develop penetration in these markets. Analysis of the potential markets for promising CATV services is badly needed but has not been undertaken extensively to date. CATV operators should consider undertaking market tests in which promising service alternatives are offered in limited areas for a reasonable period of time. Such market tests would provide information on the costs of offering these services as well as on the consumer demand for them. The investment in such research need not be large, yet the information gained might well be invaluable in making plans that promise a successful attack on major urban markets when economic conditions become favorable for such action. Execution of an aggressive, well-designed marketing strategy may then provide the desired payoff in penetration of major markets and the prospect of profit commensurate with the risk involved.

This chapter has explored the problems of and outlined the opportunities for developing CATV in the 100 major markets. There are a number of special services available over CATV which will give it a competitive edge in these markets; these will be discussed in the following chapters.

The potential of two-way interactive services will be examined in Chapter 2. A more immediate possibility, which does not require the complex, two-way transmission capability, is pay-cable television, also known as premium entertainment. Pay cable is the focus of Chapter 3.

Chapter 4 looks at origination of programs having special appeal to local viewers and ways to obtain and disseminate them. More advanced forms of program dissemination are covered in Chapter 5, including program networks, regional networks with independent TV stations, and interconnection of cable systems via microwave or satellite.

Finally, Chapter 6 reviews the development and use of the video cassette and assesses the extent to which it is likely to be a serious competitive threat to the long-run development of CATV.

NOTES

1. "A Short Course in Cable, 1974," *Broadcasting*, Apr. 22, 1974, p. 23.
2. Adapted from E. Stratford Smith, "The Emergence of CATV: A Look at the Evolution of a Revolution," *Proceedings of the IEEE* 58, No. 7 (1970): 968–70.
3. E. Stratford Smith, "The Emergence of CATV...," pp. 970–71.
4. "The FCC Delivers on Cable," *Broadcasting*, Feb. 7, 1972, pp. 17 ff.
5. Fortnightly Corp. v. United Artists Television, Inc., 392 U.S. 390 (1968).
6. CBS v. TelePrompter, Second Circuit Court of Appeals, Docket No. 72–1800, March 8, 1973, reprinted in S. R. Rifkin, *Cable Television: A Guide to Federal Regulations* (Santa Monica, Calif.: Rand Corp., R-1138-NSF, Mar. 1973).
7. "Copyright Bill Is Moving Target as It Leaves McClellan's Hands," *Broadcasting*, Apr. 15, 1974, p. 17.
8. "Trying to Keep Small Systems out of the Copyright Bank," *Broadcasting*, Dec. 31, 1973, p. 45.
9. "Copyright Bill Is Moving Target...," p. 17.
10. *The Video Publisher*, June 20, 1974, p. 1.
11. *CATV*, Aug. 12, 1974, p. 5.
12. Walter S. Baer, *Cable Television: Handbook for Decisionmaking* (Santa Monica, Calif.: Rand Corp., Feb. 1973), pp. 91 ff.
13. "Federal/State/Local Regulatory Jurisdiction," *Official Transcript*, 22d Annual NCTA Convention, Management Volume, June 1973, pp. 305–6.
14. *Ibid.*, pp. 322–23.
15. *Cable Television: Communications Medium of the Seventies* (New York: Samson Science Corporation, 1970), p. 17.
16. "Tough Test in New York: Can Wired City Be Made to Pay?" *Broadcasting*, Oct. 29, 1973, pp. 28–33.
17. *Cable Television: Takeoff into Sustained Growth* (New York: Samson Science Corp., 1972), p. 39.
18. Sloan Commission, *On the Cable* (New York: McGraw-Hill Book Co., 1971), pp. 39, 215.
19. Walter S. Baer, *Cable Television: Handbook for Decision-Making*, pp. 7–10.
20. Walter S. Baer, *Cable Television in the United States–Revolution or Evolution?* (Santa Monica, Calif.: Rand Corp., Feb. 1974), pp. 1–4.
21. U.S., Department of Commerce, *U.S. Industrial Outlook*, 1974 (Washington, D.C.: U.S. Government Printing Office), pp. 288–89.
22. *The Complete Guide to Cable Marketing*, NCTA Marketing Workshop Transcript (Dallas, Tex.: Public Affairs Department, National Cable Television Association, 1973), p. 198.
23. See Samson, *Cable Television ... Growth*, p. 39.
24. *Broadcasting*, Feb. 12, 1973, p. 77.
25. Samson, *Cable Television ... Growth*, p. 9.
26. *The Complete Guide to Cable Marketing*, pp. 202–6.
27. *Ibid.*, pp. 215–16.
28. *Ibid.*, pp. 214–15.
29. *Ibid.*, pp. 196–97.
30. "NCTA Told Cable Growth Hinges on Changes in Method of Regulation," *Broadcasting*, Apr. 29, 1974, p. 21.

31. "Pie in the Sky Turns to Egg on the Face in Big-City Cable," *Broadcasting*, Dec. 10, 1973, p. 42.
32. "Urban Cable: Still Tomorrow's Battlefield," *Cablecast* [Paul Kagan Associates, Inc.], Apr. 23, 1974, p. 6.
33. *Ibid.*
34. *Cablecast*, July 18, 1974, pp. 1–2.
35. *Ibid.*
36. *Ibid.*
37. Louis E. Boone, "The Diffusion of an Innovation: A Socio-Economic and Personality Trait Analysis of Adopters of Community Antenna Television Service," *Dissertation Abstracts, Economics* [University Microfilms, Ann Arbor, Mich.] Aug. 1969, pp. 465-A, 466-A.
38. Melvin A. Harris, "Television Consumption Behavior: Channel Use in Relation to Channel Availability," *Dissertation Abstracts, Mass Communications* [University Microfilms, Ann Arbor, Mich.], March 1972, p. 5257-A.
39. *Cablecast*, June 18, 1974, p. 4.
40. "New Revenues Bring Bright Hue to Cable's 1974 Prospects," *Broadcasting Cable Sourcebook*, 1974, p. 4.
41. "One Man's View: Cable's Going down Wrong Road," *Broadcasting*, Oct. 15, 1973, pp. 29–30.

II

TWO-WAY INTERACTIVE CABLE SERVICES

In major urban markets where three network signals and one, two, or three independent stations plus an educational station are available via rabbit ears the cable operator will find it difficult to attract subscribers. Origination of outstanding entertainment programs is one way of developing a competitive edge that might help to build penetration, but this tactic requires a substantial program investment prior to the receipt of revenues from additional subscribers and/or from sale of advertising time. Pay cable television is another possible alternative way to to offer the kind of entertainment fare that will build penetration and increase revenues from existing subscribers. Observers have also suggested that two-way interactive cable services may attract subscribers who otherwise might not be accessible as well as providing a basis for increasing revenues from existing customers and business firms who might use these facilities in their marketing efforts. Let us examine the promise of two-way interactive services from the standpoint of the cable operator who is planning profitable penetration of a major urban market under the difficult competitive conditions outlined above. The problems encountered in equipping a cable system for two-way services and in marketing such services also merit consideration.

Possible Services

CATV originally developed as a distributor of commercial television programs. In addition to transmitting such material "downstream" to the subscriber, a coaxial cable can carry information back "upstream" from subscribers to the head-end of the system.[1] This ability has encouraged cable operators and others to discuss the possibility of offering a great variety of two-way interactive cable services having potential appeal to consumers, institutions, and business.

An extensive list of interactive services which could be provided by cable was compiled by Walter S. Baer in November 1971 (see Figure 1). This list has not been sorted as to economic feasibility or social usefulness, but it suggests the wide range of possibilities from which cable operators might choose. Among the services which should be considered for individuals are:

Subscription television
Fire and burglar alarm monitoring
Remote shopping; catalog displays
Ticket sales
Banking services
Interactive instructional programs
Employment, health care, housing, welfare, and other social service information

Possible services for business include:

Utility meter reading
Industrial security
Production monitoring
Industrial training
Teleconferencing
Television ratings
Opinion polling
Market research surveys

These services, among others, may benefit government:

Televising municipal meetings and hearings
Classroom instructional activity
Education extension classes
Teleconferencing
Surveillance of public areas
Fire detection
Education for the handicapped

From the cable operator's standpoint, the list is bewildering. Which consumer services are likely to have greatest appeal thus building penetration in major markets and adding to system revenues? What are the costs of providing the various services and, accordingly, which can most easily be financed in the short run? Which business and government services have good profit potential? In what sequence should such services be offered? These questions will be considered in the pages that follow.

Technical Aspects of Interactive Cable

Since March 1972 the FCC has required each new major-market cable system to "maintain a plant having technical capacity for nonvoice return communications." The FCC explains

Services for Individuals

Interactive instructional programs
Fire and burglar alarm monitoring
Interactive TV games
Quiz shows
Subscription television
Remote shopping
Special interest group conversations
Electronic mail delivery
Electronic delivery of newspapers and periodicals
Computer time sharing
Videophone
Catalog displays
Stock market quotations
Transportation schedules
Reservation services
Ticket sales
Banking services
Inquiries from various directories
Local auction sales and swap shops
Direct opinion response on local issues
Electronic voting
Subscriber-originated programming
Interactive vocational counseling
Local ombudsman
Employment, health care, housing, welfare, and other social service information
Library reference and other information retrieval services
Dial-up video and audio libraries

Services for Business

Television ratings
Utility meter readings
Control of utility services
Opinion polling
Market research surveys
Computer data exchange
Business transactions
Credit checks
Signature and photo identification
Facsimile services
Report distribution
Industrial security
Production monitoring
Industrial training
Teleconferencing
Corporate news ticker

Services for Government

Computer data exchange
Teleconferencing
Surveillance of public areas
Fire detection
Pollution monitoring
Traffic control
Fingerprint and photograph identification
Civil defense communications
Area transmitters/receivers for mobile radio
Classroom instructional television
Education extension classes
Televising municipal meetings and hearings
Direct response on local issues
Automatic vehicle identification
Community relations programming
Safety programs
Various information retrieval services
Education for the handicapped
Drug and alcohol abuse programs

Fig. 1. Some proposed interactive services for cable television. (Source: Walter S. Baer, *Interactive Television: Prospects for Two-Way Services on Cable* [Santa Monica, Calif.: Rand Corp., Nov. 1971].)

NOTE: Not all these services are likely to be economically feasible on cable television networks. Some may not even be socially desirable. They have been compiled from various reports, FCC filings, corporate brochures, and advertising materials.

We are not now requiring cable systems to install necessary return communication devices at each subscriber terminal. ...It will be sufficient for now that each cable system be constructed with the potential of eventually providing return communication without having to engage in time-consuming and costly system rebuilding. ...When offered, activation of the return service must always be at the subscriber's option.[2]

There are different ways to achieve two-way communication on cable television systems. According to Pilnick and Baer:

The two basic technical approaches to two-way transmission are: (1) use separate cables for upstream and downstream transmission; (2) send signals in both directions simultaneously on the same cable, using different frequency bands to separate the upstream and downstream signals. ...Having a separate cable for upstream transmission presents fewer technical problems and offers more upstream capacity, but is more expensive. ...Carrying signals in both directions simultaneously on a single cable costs less than installing separate cables but is more complex.[3]

In his study of interactive television Baer classifies the many proposed new services into six broad groupings according to common technical requirements:

1. One-way broadcast services
2. One-way addressed services
3. Subscriber response services
4. Shared voice and video channels
5. Subscriber initiated services
6. Point-to-point services

After estimating subscriber equipment costs for each group, Baer concludes:

On the basis of subscriber equipment cost alone, one-way broadcast services, subscriber response services, and shared-channel services appear more feasible in this decade for mass home audiences than the other service groups. In particular, information retrieval and other subscriber-initiated services must await the development of low-cost, reliable terminals before they become attractive to home subscribers. Some difficult system design and software problems must also be overcome before subscriber-initiated services can be offered on a mass basis.

We are not concerned with one-way broadcast services at this point. One-way addressed services include the electronic delivery of mail, newspapers, magazines, and other documents. According to Baer, although a business market may exist today [November 1971] for hard-copy, addressed document delivery by cable, the cost seems too high for home subscribers over the next five years. Document recording on videotape or other soft-copy device does not appear to be an important service in itself, but it might become attractive if video-tape recorders are purchased for other reasons (such as recording of television programs for future playback). Only a small percentage of cable subscribers may be expected to have video-tape recorders in the next five years.

Two-way services requiring small quantities of return data from subscribers were evidently what the FCC had in mind in the 1972 rules for cable. According to Baer, "for these services a scanner at a central location would ask each subscriber in turn to respond to one or several queries. If the subscriber (or monitoring equipment installed at the subscriber's location) has a response, that information is sent in digital form upstream from his terminal to the central point where it is recorded or processed."[4]

These messages could include responses to questions such as whether the television receiver is turned on and, if so, to what station; questions asked by the instructor on an educational program; opinions on proposed city ordinances; and orders to buy home movie equipment displayed on the television screen. Fire and burglar alarm messages could also be sent automatically to a central station. Gas or electricity meter readings could also be transmitted automatically to a recording point.

or response over telephone

Baer explains, "Each of these queries can be answered by pushing a button or by automatically sending a few digits of information to the central location. They require, therefore, relatively low information or narrowband responses.[5] The system also could be capable of turning switches on or off remotely at the subscriber's location. This would permit the sounding of a fire alarm in the home (as well as sending an alarm to the nearest fire station), remote on-off switching of a special channel (for example, medical information for physicians), or remote on-off switching of appliances."[6]

According to Pilnick and Baer

present technology our advancement would be to add phone service.

These messages all have common characteristics. They require much less bandwidth than voice conversations, and they can be encoded in digital form for rapid computer processing. Moreover, digital messages from thousands of subscribers can be packed together into a single data stream that uses the upstream cable capacity very efficiently.

Each subscriber would have his own digital code or "address" for two-way response services. A computer at the headend (or some other suitable location) would query each subscriber in turn, using a special downstream channel. This technique is known as "polling." A two-way

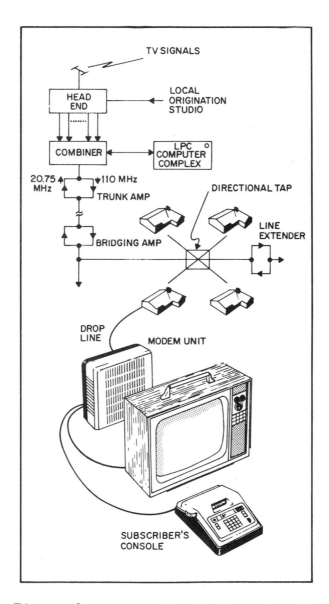

Fig. 2. **Diagram of a two-way cable television system.** (Source: *Subscriber Response System,* brochure published by Theta-Com SRS Division of Theta-Com of California, Los Angeles, copyright 1973. Reproduced with permission.)

terminal attached to the cable and television set would record the subscriber's messages, store them, and send them upstream when the terminal was polled. The messages would then be recorded at the headend computer or sent on to the city council chambers, the police station, or the department store. [See Figure 2 for a diagram of a two-way cable system.]

The basic subscriber response terminal looks like a small box with a telephone-like keyboard and a lock to prevent unauthorized use [see Figure 3]. A tunable converter might be built in, and smoke sensors, burglar alarms, and utility meters could be connected to it. Several companies are now [1973] experimenting with prototype subscriber terminals. They cost close to $1,000 today, but industry sources estimate that further development and mass production may reduce the price to $100 or so by 1980....

Other capabilities, such as browsing through a catalog displayed on the television screen, making theater reservations, or requesting a paragraph from the *Encyclopedia Britannica* will require more complex subscriber terminal equipment.

cost for phone hookup should not be prohibitive by 1990.

The Subscriber Response System is designed for two-way capability with a modest initial investment and can be expanded in a modular fashion as the number of subscribers, the traffic, and the demand for additional services increase. All without obsolescence of previously installed equipment.

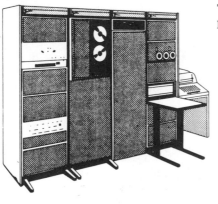

The two-way communications take place between the Local Processing Center (LPC) computer complex and the subscriber terminal which consists of a Control Console and a Modem Unit.

The LPC equipment can be located at the head end, at the local origination studio, or even remotely from the local CATV system.

Local Processing Center (LPC) Computer Complex

Fig. 3. Subscriber terminal control console, local processing center computer complex, and subscriber terminal Modem (Theta-Com Subscriber Response System). (Source: *Subscriber Response System*, brochure published by Theta-Com SRS Division of Theta-Com of California, Los Angeles, copyright 1973. Adapted with permission.)

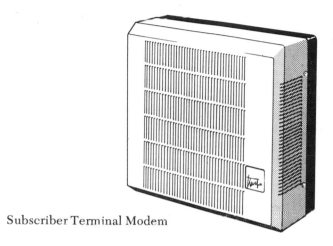

Subscriber Terminal Modem

The Modem performs most of the digital signal processing at the subscriber end and is the interface for all accessories used in the system.

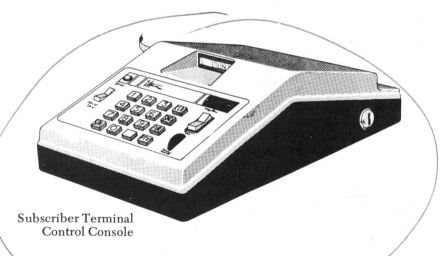

Subscriber Terminal
Control Console

The Control Console contains all operating controls, the channel selector, a keyboard and strip printer allowing the subscriber to engage in two-way communications with the Local Processing Center.

(Fig. 3, Cont.)

Institutional users of the cable system may want and be able to afford more expensive terminals and greater upstream capacity. These users include:

- Businesses that want high-speed computer data exchange
- Industrial plants that televise extension courses with student feedback to a nearby university
- Schools that want two-way video for in-class instruction and after-school teacher meetings
- Hospitals that exchange medical records and diagnostic test results and hold two-way video consultations
- Police agencies that transmit fingerprints and photos among precinct houses, or monitor streets and public areas with remote television cameras
- Local government agencies that want two-way video links for teleconferencing.

Experiments in each of these uses are under way today [1973], although they generally use microwave links, closed-circuit systems, or the telephone network rather than cable.[7]

In projecting the near-term development of new cable services, Baer summarizes as follows:

Subscriber response services, perhaps with shared voice return channels, seem ... likely candidates for home use in the next five years [from November 1971]. The investment cost for the basic two-way equipment required would amount to roughly $150–$340 per subscriber, over and above the $125 per subscriber calculated for conventional one-way cable service. Two-way services that could be provided with this equipment include audience-counting for advertisers and programmers, remote shopping, interactive entertainment and instructional programming, opinion polling, and selection of subscription or limited-access channels.

Other response services such as meter reading, fire alarm monitoring, and environmental monitoring would require additional equipment. More sophisticated and costly services such as information retrieval and computer-aided instruction could be added to the basic response system as they prove feasible.

With this capital investment and reasonable assumptions about operating costs, a cable operator would need additional monthly revenues of between $4.50 and $13.00 per subscriber to break even on two-way response services. This means doubling or tripling his present monthly revenue from one-way television distribution.

Most of the added revenue would have to come from increased monthly subscriber fees, although advertisers, business firms, utilities, schools, and government users would pay for services of benefit to them. Expected revenues from specific services cannot be estimated at the present time, since no real field experience or evidence of consumer demand is yet available.

Providing a mix of response services supported by home subscribers, business, and government users appears to be a better strategy for the cable operator than supplying a single service alone.[8]

Two-Way Demonstration Projects

Field tests of two-way cable communication began in 1971. Initially these were tests of equipment for subscriber response services, not attempts to get data on the consumer demand for two-way communication offered at a price. According to *TV Communications*, two-way systems were being tested or planned in sixteen different systems as of June 1972.[9] (See Table 6, page 42.) Only five of these tests involved video, voice, and data; these were in: Orlando, Florida (American TV & Communications); Overland Park, Kansas (Telecable Corporation); Johnathan, Minnesota (Community Information Systems); Irving, Texas (Tocom, Incorporated); and Reston, Virginia (Mitre Corporation). As might be expected, a number of technical problems have shown up in these tests. Additive upstream noise has been particularly vexing. Since the technology for two-way response services exists, however, it is anticipated that these difficulties will be resolved in time.

Mitre Corporation test in Reston, Virginia[10]

Beginning about July 1971, Mitre Corporation, supported by a $700,000 grant from the National Science Foundation, began a two-year study to develop a two-way interactive cable television system with emphasis on the educational advantages such service can provide. This demonstration involved subscriber-initiated services using a prototype frame-stopping terminal; it was conducted by Mitre Corporation on Warner Cable's Reston, Virginia, system.[11]

Since the Reston cable system did not have two-way transmission capability, the upstream link to the central computer was a Touch-Tone telephone. One home terminal in Reston was connected to a computer and a variety of services and educational concepts were tested. By dialing a specific number which connected the terminal to the computer and then pressing a certain digit on the phone, the viewer could receive a variety of preprogrammed information.

The services that were demonstrated included access to the following:

Table 6

EARLY TWO-WAY TV DEMONSTRATION PROJECTS

Location	Sponsors	Date of Project and Capability Demonstrated
Los Gatos, Calif.	TelePrompter Hughes Aircraft Corp. Fairchild Instrument Corp.	June 1971— transmission. tests only
El Segundo, Calif.	TelePrompter Hughes Aircraft Corp	Jan. 1972—prin- cipally data
Orlando, Fla.	American TV & Communications Electronic Industrial Engineering	May 1972— video, voice, and data
Pensacola, Fla.	Advanced Research Corp.	Sept. 1971—data
DeKalb County, Ga.	Advanced Research Corp.	June 1972—data
Monroe, Ga.	Scientific-Atlanta L.E.A.A.	Future—data
Overland Park, Kansas	Telecable Corp. Electronic Industrial Engineering Vicom Manufacturing Co.	June 1971— video, voice, and data
Carpentersville and Crystal Lake, Ill.	LVO Cable and Scientific-Atlanta, Oak Electro/Netics	Near future— data
Dennis Port, Mass.	Rediffusion, Ltd.	1971—data
Jonathan, Minn.	Dept. of Housing and Urban Development Community Information Systems with General Electric help	Feb. 1972— video, voice, and data
New York City, N.Y.	Sterling Communications Video Information Systems	Feb. 1971—data
Brooklyn, N.Y.	Goldmark Comm. Warner Communications, Bedford-Stuyvesant	Future—video, voice, and data "showpiece"
Akron, Ohio	Television Communications	June 1971—trans- mission tests only
Irving, Texas	TOCOM, Inc.	1972—video, voice, and data
Brigham Young University, Provo, Utah (actually CCTV)	BYU, Cascade Electronics, Ltd. Hammett and Edison	1970—video
Reston, Va.	Continental Transmission Co. MITRE Corporation	July 1971— "frame stopping" terminal; respons€ via telephone

Source: *TV Communications*, June 1972. Reprinted by permission. For more information contact TV Communications, 1900 W. Yale, Englewood, Colorado 80110 (303) 761-3770.

1. Calculations—e.g., extracting the square root of a number
2. Community information
3. A "Dutch auction," which offered various products for sale
4. Community bulletin board
5. Medical information
6. Payment for products or services via cable—e.g., payment for magazine subscription where the amount is deducted from the subscriber's bank balance
7. Announcement of employment opportunities—e.g., information on summer jobs for teenagers
8. Instructional program—e.g., lesson in grammar, involving viewer's answering questions asked in writing on the screen
9. Drill in mathematics—again, where viewer answers questions written on the screen
10. Games—e.g., "pick up sticks"
11. Retrieval of video tape on topic of interest to viewer—voice plus picture

In the demonstration the question was asked as to how many viewers would be willing to purchase some predetermined combination of the above services. Mitre officials noted that the cost of providing such services might run approximately two-thirds or less than the cost of regular telephone service (possibly $14 per month). In addition to revenues from subscribers, it was noted, the cable operator might also get income from leasing channels to programmers of pay entertainment, to business firms, and others. What services viewers would be willing to pay for and how much they would be willing to pay per month could only be determined by market tests at some point in the future when two-way interactive services would be offered on a commercial basis.

Mitre's operational experiment in two-way cable[12]

After the technical feasibility of Mitre's two-way system was verified in the 1971 Reston test, the organization proposed to the National Science Foundation that the experiment be expanded to 40 subscribers on the Reston cable system using the Touch-Tone telephone to connect viewers with the central computer. Two major drawbacks to the proposed project were noted: (1) Reston's cable system was one-way, and it was believed that the next test should be on a two-way system; (2) analysis of the demographics of subscribers on the Reston system indicated that they were atypical; a more representative community was thought to be desirable for further experimentation.

Accordingly, in June 1973, Mitre officials developed a proposal for a new project, an operational experiment in two-way interactive cable television. This second phase of Mitre's study is expected to take about three and one-half years. The first six months will be used to develop the programming to go on the system, arrange with a cable system to participate in the project, and select an equipment manufacturer from which to procure the necessary two-way hardware. It was estimated that it will take a year to install the system; the remainder of the time will be for experimentation.

Mitre Corporation officials requested a grant from the National Science Foundation to fund the project, its purpose being to determine the educational advantages of bi-directional cable.[13] In the test cable subscribers will be able to select from a computer bank containing numerous formal audio-visual educational materials, as well as community information and certain computerized educational games.

After screening 39 communities as possible test sites, the list was narrowed down to four: Stockton, California; Akron, Ohio; Peoria, Illinois; and Spartanburg, South Carolina. In February 1974 it was announced that Stockton had been chosen on the basis of its desirable educational, social, and economic diversity. The 1970 Census showed a population of 115,000 in the metropolitan Stockton area. Agriculture and shipping provide the chief employment in the community with only a small amount of industry there.

The community's cable system was being built by the Big Valley CATV system (a subsidiary of Continental Cablevision), which was installing a two-way dual cable system with six separate service areas trunked back to a central point. Four areas were scheduled to have 30 channels; two areas, 20 channels. Mitre can use 15 channels in its experimental work.

According to Big Valley's plans, by 1975 the cable system was to pass by 54,000 homes; 35–50 percent penetration was projected by that time, which meant about 20,000 to 27,000 CATV subscribers by 1975. (2,000 subscribers had CATV service in February 1974, but construction was proceeding rapidly.) Because of the manner in which the system was planned, it could be broken into six segments based upon ethnic distribution of the pupils in schools serving the city.

Under the proposed plan Continental Cablevision is to cover the cost of building the dual cable system which will permit two-way communication (originating either downstream or upstream). Mitre Corporation originally suggested the installation of 1,000 terminals in homes of volunteers selected from among cable subscribers who meet criteria to be established after a survey is made of the Stockton metropolitan area. It was originally estimated that the experiment would cost $3.2 million during the first three years. This included the investment in terminals and necessary programming but not the cost of the cable system (to be covered by Continental Cablevision).

The planners expected that it would be nine months from the start of the project before all hardware was on line and running, and they expected that a year would be required to "debug" the system. It was not planned to charge subscribers for two-way services initially. At some point in the future, after an adequate user base has been achieved, two-way service charges would be levied.

Some of the questions to be studied during this experiment include the following: (1) What educational services should be offered? (2) How can they best be furnished? (3) What program content should be provided? (4) What should be the style of delivery of program material?

According to the proposal, the Educational Testing Service is to evaluate the success of the project as it progresses. This will include an appraisal of Mitre's ability to develop appropriate program content and an evaluation of the effectiveness of alternative styles of delivery. Experience gained by Mitre will provide the basis for a publication of a technical and economic analysis of two-way interactive television with special attention to the effectiveness and attractiveness of interactive television as a means of providing needed educational services. As indicated above, the results of the Stockton experiment are not likely to be available for at least 3½ years from the time the project actually gets under way.

Subsequent developments

At the time the Stockton operational experiment was described at the University of Michigan Telecommunications Faculty Seminar in February 1974, Mitre officials hoped the National Science Foundation (NSF) would make the proposed grant and that the project would get under way in about six to eight weeks. This expectation was not met, however, since the NSF did not fund the project in the amount originally proposed. The final NSF decision was not expected by Mitre officials before October 1975.

There also have been some revisions in cost estimates for the Stockton operational experiment. As of June 1975, the cost of the hardware for the Stockton two-way system was estimated to be $1,000,000. Mitre officials now plan to install 450 subscriber terminals with 20-character keyboards—although the system as planned would be capable of supporting many more terminals. The actual number that eventually would be installed on the system, unknown at present, will depend strongly upon subscriber usage, which is the major unknown to be explored by the project. The cost of the first 450 subscriber terminals is included in the $1,000,000 estimate for hardware. It is estimated that the incremental cost of adding a single terminal will be around $250. This would provide a pay-TV terminal which costs much less than quotations received on subscriber terminals used in developing the original plans.

This type of service could be delivered to the subscriber at a cost ranging from $10 to $20 per month using today's technology, a Mitre spokesman estimates. Measured in terms of the time a subscriber would be using the two-way system, the cost is estimated at two to five cents per contact-minute. Moreover, this figure is surprisingly independent of the amount of subscriber usage.

The new target date for beginning operations of the Stockton experiment is October 1976, although according to the spokesman this date is far from firm because of the uncertainty about NSF funding. Except for the modifications mentioned above, however, he indicated that the description of the proposed Stockton experiment provided here is still generally correct.[14]

While it is clear that the Stockton experiment, if implemented, will be concerned with the social impact of interactive television as applied to educational uses, at a later point the project may be expanded to add consumer education as well as health-care services. The Stockton experiment is not concerned with testing the appeal of various other two-way services, such as premium entertainment, shopping service, and checkless banking transactions. Continental Cablevision, through the Big Valley CATV system, would appear to have an excellent opportunity to explore some of these possibilities on its own behalf, however.

Jonathan/Chaska Community Information Systems project

In February 1972, Community Information Systems (CIS) undertook a pilot two-way cable demonstration in Jonathan Village/ Chaska, Minnesota, under a $175,000 grant from the U.S. Department of Housing and Urban Development.[15] Community Information Systems, Incorporated, is a private business engaged in the interactive television field which provides consulting services, equipment and/or management assistance. The demonstration facility and the associated cable system in Jonathan/Chaska were built by CIS, under license from General Electric Company for patents and technical expertise.

The communities of Jonathan Village, a new planned community, and Chaska proper together constitute the political unit of Chaska City, located in Carver County, Minnesota, about 25 miles southwest of Minneapolis. As of October 1972 the population of the total community had reached 5,398, of which about 1,650 people were located in Jonathan Village, the first new town to obtain loan guarantees under Title IV of the 1968 Housing Act. In the past Chaska City had been a small urban center servicing a predominantly agricultural area. With the high growth rate of Jonathan Village in the two years prior to 1973, Chaska City was in the process of becoming a residential area tied socially and economically to the Minneapolis–St. Paul metropolitan area. It was anticipated that within the next 20 years the Jonathan Village would provide housing, employment and services for 50,000 people.

The Jonathan/Chaska experiments are especially interesting because, along with a demonstration of key two-way interactive cable services and a cost estimate of providing them, they provide an estimate of the demand for such services, secured by means of survey research techniques.

The CIS demonstration center was designed to inform the public about two-way interactive cable communications. The equipment arrangement includes a "hands-on" area and a home communications center area; this makes it possible to inform people of many potential services. Visitors to the demonstration center are given an audio-visual presentation, which deals with how CATV started and its initial purpose of providing better TV reception and then explains what cable can provide today in terms of local origination and increased channel capacity. The viewer is then introduced to the Community Information System and, through a series of examples, he learns how two-way cable can be used in areas such as interactive education, information retrieval, protection services, etc.

The hands-on station gives visitors practical experience in operating interactive terminals to receive services. The visitor can select any one of the following interactive programs: information retrieval, interactive education, interactive entertainment, point-to-point data communications, protection services, merchandising, and opinion polling. Equipment available for consumer use includes a television set, an input/output typewriter (to print out messages and send alphanumeric messages), and a Responsor℠ terminal (used for entering numbers to "call up" specific programs and information). It also transmits alarm messages (fire, intrusion) to the Communications Center.

Using the terminal, the viewers enter the proper access codes for the selected program or service. The selected program is then presented on a color television set. To provide realism, a home information center adjacent to the hands-on area simulates a present-day den or family room setting containing a television set and a Responsor™ terminal. Visitors are encouraged to operate the equipment and ask any questions they may have.

As a means of getting information upon which to base an analysis of the demand behavior of consumers, a survey was taken of a sample of households in Jonathan/Chaska during an 18-day period beginning in October 1972. The sample of households in which interviews were conducted included 292 in Chaska proper and 120 in Jonathan Village, or a total of 412. Of the 412 included in the sample, 40 did not respond, leaving a total of 372 completed interviews.

An important part of the survey's design was to inform the community, and in particular the respondent households in the sample, about the Community Information System and its services in advance of the survey. Of the 372 households interviewed, 16.4 percent visited the demonstration, 22.8 percent discussed it with friends, 38.2 percent read a brochure describing it, while an additional 20.3 percent read some of the brochure.

Although a CATV system had not been constructed in Jonathan/Chaska as of August 1973, there was a two-way cable link up installed and working at Jonathan that was serving as a test bed for the capacity to assist in health delivery and education. The network connects two medical clinics with a community hospital to assist the doctors, paramedics, and patients in the community. The other link connects the local high school and grade school to the CIS Information center allowing the student to dial television programs for display in any classroom. The network also allows two-way audio-visual communication between students in their classroom and local business and professionals at the Village Center in Jonathan.[16]

The report of the Jonathan/Chaska Community Information System experiments summarizes their results as follows:

- The demand for basic cable service was indicated to be 64.8 percent in the Jonathan Community and 30.3 percent in Chaska. Overall demand was 40.3 percent.

- Over 95 percent of the respondents interested in the basic CATV service were also interested in one or more extra services before prices were introduced.

- At the low price range ($4 to $6 per month per extra service), 83.3 percent of the prospective CATV subscribers desired one or more extra services. On the average, each extra service subscriber demanded 2.1 extra services at an average monthly cost of $10.00 at the low price range.

- As would be expected, demand for extra services decreases with price. At the median and high price categories ($5.75 to $9.75 and $10 to $14.25) approximately 58 percent and 30 percent of the prospective basic CATV subscribers, respectively, still demanded one or more extra services.

- Several generalizations can be made about the demand and its structure and elasticity for extra services. It is clear that the most popular extra services among prospective CATV subscribers in the area were
  —Education and learning
  —Premium programs
  —Alarms
  These services are of great interest to all segments of the population, regardless of background characteristics or community of residence. Moreover, the demand for these popular services tends to be relatively inelastic—i.e., as prices change the demand remains fairly constant. The other services, especially Information Please and Home Shopping, have less but nonetheless an appreciable demand. Their demand schedules also tend to be more elastic than those for the other services.

- In household composition and family income, the population of the total sample representing the communities of Chaska and Jonathan combined is similar to the U.S. national average. In educational attainment and occupational status, the families sampled more closely resemble the population of the Minneapolis-St. Paul SMSA. The results of this program can be confidently generalized to many metropolitan areas.

- It is important to note that family income is not a principal determinant of demand levels and elasticity. Rather, the interrelated complex of variables such as age, children in the home and their ages, stage in the life cycle, and life styles seem to be more important than income alone. Income seems to be only a limiting factor, becoming a real constraint in the lowest income bracket and reflecting easy consumer indulgence at the higher income brackets.

- Community Information Systems has constructed and now operates a 14-mile interactive two-way experimental plant in and around the "new town" of Jonathan. It is expected that the size and scope of this test bed will be continually expanded over the next several years. Configured as a modern, conventional system with extra service overlays, it is envisioned that this plant will be the basis for an area-wide, innovative community information system.

- A plan for the development and implementation of an economically viable extra service business utilizing the area-wide community information system is presented. Projections based on results of the community survey have been utilized to structure an extra service package.

This package would consist of four services—Education and Learning, Premium Programs, Home Security, and TV Games—and would be priced at $22.50/month.

Revenue and profit projections for the regional Community Information System show that if the marketability of extra services can be established, the revenue and net profit potentials of the community information system at maturity (10 years) are more than twice and three times, respectively, those of a conventional CATV system.

- As a direct result of HUD's initiative, additional projects have been funded in health delivery and educational applications in which the community is actively involved as participants. This high level of enthusiasm, participation, and acceptance exhibited by the community can contribute significantly to the future success of such programs and demonstrates the leverageable impact of sponsored research.

- Further demonstration and evaluation programs are needed to measure not only operating characteristics and financial feasibility but also to determine the sociocultural implications of interactive wideband communications.

Program summaries of services that show near-term potential of becoming economically self-supporting, prepared with the aid of community and area residents and organizations, are presented for education, telemedicine, and municipal services.

In the same report on the Jonathan/Chaska Community Information System Experiments, CIS outlined the following long-range goals for Phases II and III of this program:

Provide the U.S. Department of HUD with critical information and operating data to be used to guide the development of cable communication in all communities, allowing the Department to:

—Use the results of the Phase I project to initiate full-scale demonstration projects in a number of communities which could include Chaska and Jonathan

—Determine the utility and cost of the services Chaska/Jonathan residents prefer, using test and evaluation programs which employ the full-scale demonstration CIS

—Evaluate alternate methods of financing the implementation and operation of the CIS and the various services it can provide

—Test various combinations of public and private participation and management and try various methods and models for performing the functions essential to the provision of services.[17]

In an interview a spokesman for CIS indicated that the next step should be actually implementing two-way interactive cable television—i.e., Jonathan/Chaska would franchise a CATV system offering not only transmission of over-the-air broadcast signals but also extra services found to be of special interest to consumers in the survey described above. This step would permit experimentation to determine consumer demand for extra two-way services for which there would be a charge—especially important because responses to survey questions on what prices respondents would pay for various services tend to be inflated.

Alternative levels of prices could also be tested, as well as alternative marketing strategies for promoting the purchase of extra two-way services by subscribers.

It was reported that large retailers were exploring the possibility of conducting shopping experiments on the Jonathan/Chaska TV system when it became operational. Certainly, it would be highly desirable for market testing two-way interactive cable television along the lines described above. Results of such experimentation would provide valuable information for CATV operators considering the implementation of two-way cable services on their own systems.

Subsequent developments

A telephone conversation on May 21, 1975, with E.D. McCormick, President, Community Information Systems, Incorporated, provides the following information on developments since August 1973:

Although Phase I experimentation produced evidence of the potential value of two-way interactive cable services, CIS has not yet been able to carry out proposed additional tests in actual market conditions as proposed above. No action has been taken on the proposal that a CATV system be franchised and constructed in the Jonathan/Chaska service area. There were two key impediments to progress. (1) The high interest rates of 1974 followed by the business recession produced a tight money economy. This made it difficult to attract the venture capital needed to finance the construction. (2) The founder of the firm that developed Jonathan/Chaska passed away about 18 months ago. Negotiations for selling the corporation have not yet been concluded. Until this transaction is completed, the proposal to construct a CATV system in Jonathan/Chaska is not likely to receive the necessary consideration.

The two legs of the interactive cable systems used as a test bed for assisting health care and education are still in existence. In fact, the school leg has been extended two miles to pick up two more facilities—the junior high school in Chaska and the headquarters of the ecology instructional program in the Chaska School District.

The experimentation which continued during 1973–74 and 1974 –75 in the use of two-way interactive cable to assist in the educational process and in the evaluation of its contribution indicates, on preliminary evaluation, that two-way appears to offer some economies as compared with traditional instructional methods. But significant differences have not been found between opportunities for learning by direct contact instruction and instruction using two-way interactive audio-visual cable.

The experiment in using two-way audio-visual communication via cable in operating a rural health-care system has been fully evaluated and a report of the findings has been submitted to the Department of Health, Education, and Welfare. The evaluation checked the satisfaction of both patients and physicians with the two-way system. Patients were well satisfied; physicians reported that two-way cable did not make a significant improvement in the level of their practice or the efficiency of their work. It is believed that a fully automatic system might improve efficiency, but CIS has not proposed such an installation because of the financial investment required.

As a result of this evaluation, Lakeview Clinic is presently embarked on another study looking at patient blockages principally caused by communication and how the use of two-way interactive cable can have an impact on these blockages in the health care system.

In short, the two-way interactive cable link-ups in Jonathan/ Chaska and their associated activities are still in operation, although construction of the two-way cable system and the proposed experimentation with interactive customer services await a more favorable economic climate.

Other demonstration projects

El Segundo pilot project.[18] An interesting pilot project began in September 1972, when Theta-Com of California designed and installed a complete two-way interactive system in El Segundo, California, where TelePrompTer had constructed a two-way cable system. The project used the Subscriber Response System equipment developed by Hughes Aircraft Company and produced by Theta-Com.

According to Robert Behringer, then president of Theta-Com, the El Segundo project was undertaken to measure the market which company officials believed existed for services, data retrieval, and entertainment. Initially Theta-Com installed fourteen Subscriber Response Terminals (SRT) in El Segundo for test and demonstration purposes. This number was later to be expanded to twenty-five. A two-cable system was installed to connect a Local Processing Center computer complex to the terminal in the subscriber's home. Tests and demonstrations were to be continued until 1,000 subscribers' terminals were installed at a cost of approximately $200 each. At that point plans were to begin the delivery of actual service to the 1,000 homes. After one year of actual market testing, results of the research would be published.

During the NCTA Convention in June 1973, a wide range of services were demonstrated at the Hacienda Motel in El Segundo. They included premium TV (movies, sports events), shop-at-home services (supermarkets, department stores, ticket services), emergency alarms (fire, intrusion), and emergency medical information and services, among others.

According to Behringer, it is estimated that the cable operator could realize an average of $10 per month of added income per installation from the two-way services. The estimate assumes premium or pay television as a basis with other two-way services adding incremental income. The figures indicate $120 per year of cash flow, which would more than pay the carrying charges on the added investments. On the basis of Theta-Com estimates, it appears that the capital investment, when the units are in production, would be $200 to $250 for each home terminal equipment set. This figure includes a pro-rata share of the computer and the computer software for the head end. It does not include the cost of converting the cable system to two-way.[19]

As is usually true with developmental work on innovations, the original time schedule for this pilot project has been modified because of problems encountered in getting the two-way system operational. Original plans called for subscriber services to commence in late 1973, when it was expected that 200 SRS terminals would be installed; 1,000 terminals were scheduled to be installed and operational by January 1974. In a May 1973 conversation, however, Behringer reported that the computer had been installed and that Theta-Com engineers were debugging the system at that time. Only 14 terminals had by then been installed and these were distributed in an apartment used for test purchases, private homes, and the head end of the cable system. Behringer estimated that it would probably be six months before the experiment would be fully under way. A year later results on consumer demand for two-way services under test market conditions were to be available.

Subsequent Developments in Two-Way Projects

In the two years since June 1973, growth in the cable television industry has slowed. Such firms have experienced difficulties in

obtaining adequate capital for expansion of their systems, not to
mention capital for the installation of two-way facilities. Symptomatic of the tight money situation that developed was the sharp
rise in the prime interest rate from 6 percent in early 1973 to 10
percent by December 1973, and 12 percent by mid-1974. During
the same period there was an inflationary rise in the cost of
building new CATV systems which discouraged new construction
and investment in two-way equipment.

Under these circumstances, the following report from a January
1975 issue of *The Video Publisher* is not surprising: "Theta Cable,
jointly owned by Hughes Aircraft and TelePrompter, pulled back
from its ambitious test of two-way subscriber wideband.... The
curtailment is symptomatic of the economic state of the CATV
industry in general."[20]

H. R. Goodman, manager of the Multiplex System Program,
Hughes Aircraft Company, provided additional information on the
status of the El Segundo Pilot Project in an interview on May 29,
1975. He indicated that product line responsibility for that activity
had been transferred back to Hughes, Theta-Com's parent company.

While a total of forty-five subscriber terminals had been installed, these were used solely for engineering experimentation
and product planning study. The firm was interested in measuring
certain parameters related to long-term equipment operation on a
two-way cable system. After getting the desired data, it dropped
further experimentation.

The project was never extended to permit marketing testing of
the subscriber demand for two-way services in 1,000 homes as
originally planned. Studies of the potential demand of CATV
operators for the subscriber response terminals led company officials to the conclusion that such equipment was not what the
CATV industry was able to buy in the near future, and accordingly
it did not seem to be worthwhile for the firm to build and install
the 1,000 terminals for market testing experiments.

Mr. Goodman emphasized the point that Hughes Aircraft has
continued the development and testing of two-way interactive
equipment. The firm is, however, stressing the production of
equipment which has a more immediate prospect of return than
did terminals developed for the El Segundo marketing experiment.[21]

Orlando, Florida, test[22]

Another experiment involving the two-way transmission of
video, voice, and data was begun in May 1972 in the Orlando,

Florida, CATV system owned by American TV and Communications with equipment developed jointly by ATC and Electronic Industrial Engineering, Incorporated (EIE), a wholly owned subsidiary of RCA. According to an EIE spokesman, Phase I of this experiment involved a pilot installation of twenty-four terminals located in homes, businesses, and educational facilities. This Polycom™ system was connected to a central computer and involved tests of a number of services. One terminal was located in a local service station for use in credit card sales of gasoline, oil, and other products; billing information was picked up by the computer and made available to the service station management; terminals in subscribers' homes offered as many as three CATV channels for use in so-called narrow-casting. The terminal had a 13-digit key pad that could be used in opinion polling or in home merchandising of products and services. The company demonstrated two-way capability for use in security systems (burglar alarm, fire alarm). Two-way surveillance terminals were installed for demonstrations over one of the local freeways for traffic control. Through the terminal in the home the system had the ability to connect or disconnect service to the television set automatically. The system also had the basic capability for reading meters along with monitoring channels (i.e., recording what channels the CATV subscriber watches during a 24-hour period).

The main purpose of the experiment was to prove technical feasibility of the two-way interactive system and equipment. The experiment was judged successful on this basis.

Subsequent developments. After technical evaluation of the system was complete, the equipment was removed from operation. A product version of the system was not developed because of concern that the market for two-way services had not yet begun to evolve.

As noted, the impact of tight money followed by the recession of 1974–75 created financial problems that made cable operators reluctant to invest in costly two-way equipment. Although EIE had made a substantial investment in the development of its Polycom two-way equipment and had developed a very sophisticated system, it became evident that there was little demand for it during 1973–75.

As of mid-1975 the immediate outlook for the installation of two-way systems is discouraging. According to the EIE spokesman, a lot of CATV systems are marginal and a good many of them are losing money. Until this situation changes, they have no incentive to invest in a sophisticated two-way system. Instead, the center of interest in 1975 is pay TV, which, with the advent of satellite distribution of program material, is becoming viable. But

CATV operators are looking for the most economical equipment that will provide satisfactory pay TV operation. Consequently, many are installing equipment using a scrambler, which is less costly than a sophisticated two-way subscriber terminal.

When will a demand for sophisticated two-way equipment develop? According to the EIE spokesman, not until the economy turns around and the climate favors long-range investments of this sort. In addition, once Multiple System Operators start accumulating money from their pay TV operations, they must start thinking about other types of additional services that might appeal to subscribers and generate more revenue. When the anticipated revenue from two-way services bears a favorable relationship to the costs of installing such a system, then decision makers may consider its purchase.[23]

Irving, Texas, experiment

As of May 1973 Tocom, Incorporated, of Dallas, Texas, was in the process of installing the equipment for an experiment in two-way cable television in Irving, Texas.[24] Tocom was franchised to establish the experimental system by the Irving city council. The cable system was later sold to Leacom of Irving, a subsidiary of the C.H. Leavell Company of El Paso. Leacom is thus the operating company. According to a Tocom spokesman, the experiment was to be operated as a mini-pilot study with 20 remote units (subscriber terminals) until the system received a certificate of compliance. The initial experiment would test the technical capabilities of the two-way system; later the feasibility of offering selected two-way services would be studied. Eventually 2,000 homes were to be included in the two-way experiment.

The Tocom system includes three primary elements: The first element, a so-called remote unit, is placed in the subscriber's home and is a combination of a 26-channel TV converter and a digital transmitter-receiver, all housed in one attractive cabinet. The unit sits on top of the television receiver; each home has its own unique identification and responds with a digitally coded signal when interrogated by the central data terminal. The second element in the system is a computer-controlled central data terminal which is capable of interrogating, receiving responses from, and acting on the responses of 60,000 remote units every six seconds. The third element is a Bi-Directional Cable Distribution System, which may be of either single or dual trunk configuration, with a forward transmission bandwidth to allow for 26 channels of TV reception. The system also exhibits a reverse transmission bandwidth in the 5–25 MHz range.

The capabilities of the system to be tested include a number of possible services:

1. Using the response buttons on the remote unit, the subscriber may actively participate in such programs as home shopping and opinion polls.
2. In the same way the subscriber may participate in two-way education programs in which the home viewers may communicate with a teacher.
3. Each remote unit will accept and relay to the central data terminal three separate alarm conditions—fire, burglary, and a need for emergency assistance.
4. One or more of the 26 channels may be delegated as pay TV channels. Insertion of a key into the remote unit allows viewing of the pay TV program, and also starts the computerized billing procedure, which may be based on a fixed price per program or on the length of viewing time.
5. Other services available to the subscriber include meter reading and remote control of lights, sprinkler systems, motors, alarms, etc.
6. Upon command by the CATV operator, the Central Data Terminal will furnish a print-out listing how many TV sets are turned on and how many are watching each channel every six seconds.

The purpose of the experiment is to get data upon which to base estimates as to the size of the market for the various two-way services and to serve as a guide for deciding in which sequence two-way services should be introduced to subscribers under competitive market conditions. While the remote unit is expected to cost about $150 when it is produced in volume, Tocom executives have not yet decided how much such a system would cost subscribers when it is offered commercially after the experiment. The remote unit will be installed without charge during the test. Also, during the experiment, Irving subscribers will be charged only for the two-way services that they actually use. Cable operators in the future will have to make the decision as to whether to sell the remote unit to subscribers, lease it, or add its cost into the charges made for the two-way services utilized by the consumer.

Subsequent developments. An interview with a Tocom spokesman on May 30, 1975, provides the information that the Irving experiment was never extended beyond the original pilot study of twenty remote two-way units. In part, the experiment had been planned as a research and development activity in Tocom's own back yard to test the technical capabilities of the firm's two-way communication system. Although benefits were gained from this experience, the slump in the CATV industry discouraged Tocom officials from implementing their original plan to cooperate with Leacom by installing two-way equipment in up to 2,000 homes to test subscriber demand for interactive CATV services. In this regard Tocom executives reached the same decision as other leading producers of two-way interactive communication systems such as Theta-Com, Jerrold, Scientific Atlanta, and Electric Industrial Engineering.

The Tocom spokesman reports that CATV operators have shown very little interest in the purchase of bi-directional systems except for equipment necessary to provide pay TV service. Tocom's system would be too elaborate, and too costly, for pay TV service alone.[25]

Tocom installations in housing developments

Tocom has found a promising market for its two-way systems among the builders of housing developments, however. In May 1973, a Tocom spokesman reported that the firm had signed a $3 million contract with Rossmore Corporation to install the Tocom II two-way cable monitoring system in 11,000 homes to be constructed in a new community for people over 55 being developed in Mesa, Arizona. Deliveries to the Mesa project were to begin in December 1973, and the installation was to be complete by October or November 1974.

The contract calls for installation of the full computer and cable system in each of the 7,000 residences being built and for a projected increase of 4,000 units. Should the residences exceed 11,000, Tocom's contract will increase to $5 million. The system will provide capability for home shopping, pay TV, opinion polling, fire-burglar-emergency assistance alarms, meter reading, and remote control functions.

This installation offers the possibility of gathering additional information on consumer demand for two-way services, if the Rossmore Corporation is willing to cooperate in such a study.

Developments, 1973–74. Because of conditions in the economy since 1973, the sale of housing units in Rossmore's Mesa development has been slower than originally anticipated. Nevertheless, the Tocom II two-way monitoring system had been installed in 380 homes by mid-1975. Currently, residents are provided with intrusion, fire, and emergency assistance alarms in addition to regular cable service. As more homes are built and sold, Tocom will continue to install their two-way equipment under the terms of their contract with Rossmore. The cost of the system is added to the price Rossmore charges for the lot and the living unit.

The system includes a bi-directional cable using Tocom remote terminals in each home (i.e., a remote receiver/transmitter on each set). The computer is located in the Rossmore main office. In addition to the alarm system, remote teletype equipment provides hardcopy printouts for those who must take action on alarms— these units are located in the fire station, security guard office, police station, and nurses quarters.

As of June 1975 the system is reported to be up and running. Since it became operational, the medical assistance request switch has brought help to two residents who suffered heart attacks. One fire alarm brought help in time to save a house. There have been no intrusion reports. (Rossmore provides excellent security protection for the Mesa project.) Residents in the development are pleased with the protection the two-way monitoring system provides. Rossmore management is very happy with the installation. Indeed, the firm has ordered another Tocom monitoring system for installation at the Leisure World development under construction in Coconut Creek, Florida.

Tocom has also contracted to build similar two-way monitoring systems in other housing developments being built under Title VII HUD guarantees. Such arrangements have been made with builders in Woodlands, Texas, located outside Houston, where 25,000 homes are planned (by June 1975, several hundred units had been constructed), and Flower Mount, Texas, where 20–25,000 homes are planned (eight houses have been occupied to date). In each case the system is bought by the developer and the cost is pro-rated to the price charged for the lot and house. In a 10,000 home project, the pro-rated cost per house for the initial investment in Tocom two-way equipment is about $400 to $500 including the bi-directional cable. Developers are emphasizing the two-way monitoring systems as a special feature in promoting the sale of houses in their projects.

In both of the contracts mentioned above Tocom personnel not only install the two-way system, but they also service and manage it thereafter. This arrangement provides management with feedback as to the technical operation of its system in actual practice, the cost of system operation, and the reactions of consumers to the various two-way services made available.

In short, it appears that Tocom management has found a significant market for its two-way equipment among housing developers at a time when CATV operators are not buying such systems. Such contracts provide immediate revenue to support further research and development. Experience is being gained in the operation and management of two-way systems. Technical problems may be identified and action taken to correct them. Much may be learned about reactions of consumers as they operate the sophisticated two-way equipment. Thus, there is an opportunity to assess the demand for the various two-way services that are offered.

In the long run, however, Tocom management recognizes that if the industry is to survive it will be necessary to develop the market for two-way systems among CATV operators in urban markets. Nevertheless, selling two-way installations to housing project developers appears to be a promising way to bridge the gap until CATV operators realize there is a potential profit from providing subscribers with two-way services and begin investing in such systems.

Goldmark's ten-city study in Connecticut

One of the most far-reaching proposals involving two-way cable television has been advanced by Dr. Peter Goldmark, former President of CBS Laboratories.[26] In 1972 it was announced that his firm, Goldmark Communications Corporation, would study the broad uses of communications technology in a ten-city area in Connecticut. The firm planned to investigate the potential use of broadband communication devices in service industries, for expanding health services through mobile teleclinics, for interconnecting large educational institutions with satellite campuses, and for bringing entertainment sources into rural areas. The study, funded with a $400,000 grant from HUD and the National Sciences Foundation, is being conducted in cooperation with Fairfield University, Fairfield, Connecticut, under the title "The New Rural Society." (Goldmark is a visiting professor at Fairfield.)

In December 1974, HUD awarded Fairfield University another $300,000 to continue the New Rural Society Project. This phase of the project is to involve development of new broadband communication techniques to facilitate the decentralization and relocation of state agencies and business operations.[27]

Additional noteworthy two-way experiments

Other two-way experiments funded by government agencies involving delivery of social services include the following:

In June 1974, NSF gave a grant of $99,129 to Cable TV Information Center to test costs and benefits of two-way cable for delivery of social services in an urban setting. This project is to explore whether CATV can improve delivery of social services to the elderly and create alternatives to the present institutions which serve this increasingly isolated segment of the urban population. It will also explore whether two-way can improve a community's sense of participation in government. The test site is Peoria, Illinois.[28]

In May 1975, HEW's Bureau for Handicapped Children funded

an experiment in which Mitre Corporation is to work with the New York Education Department to install a 100-terminal, two-way system in Amherst, New York, a Mitre spokesman reports. The system will begin service in September 1975 and will experiment with providing instruction to homebound handicapped children. The two-way system will use telephone return and full keyboard terminals in the home.[29]

In May 1975, NSF also let three contracts for two-way experiments involving the provision of social services to the public. Each involves a CATV system with two-way capability in a city with a cooperative local government. These include (1) Rand Corporation's Washington, D.C., office to work in Spartanburg, South Carolina—$1.1 million; (2) Michigan State University to work in Rockford, Illinois—$400,000; (3) Alternate Media Center, New York University, to work in Reading, Pennsylvania— $200,000.

These experiments all focus on delivery of social services and will serve to explore and develop interesting uses of two-way interactive CATV. They are not expected to lead to commercially viable use of two-way, however,

Experiments in Marketing Goods and Services via CATV

Several of the two-way cable tests described above have demonstrated the possibility of offering subscribers the service of shopping in their own homes. The combination of the two-way cable, the subscriber terminal (or remote unit), and the computer-controlled central data terminal at the head end make interaction between the subscriber and the marketer possible. The widespread use of credit cards is another important facilitating factor. Before cable operators make the investment necessary to provide shopping service via this two-way interactive system, however, it is necessary for them to have some assurance that consumer demand for this service will be large enough to promise a satisfactory prospect of profit.

It is extremely significant, therefore, that several firms have taken initial steps to experiment with various versions of the electronic shopping concept. Three tests—in Toronto, Canada, in Louisville, Kentucky, and originating in New York City—have proved to be particularly interesting.

Simpsons-Sears and IBM experiment

In October 1973, for example, the *Detroit Free Press* reported that:

Simpsons-Sears, Ltd., and IBM of Canada, Ltd., said they are experimenting with a computerized catalog-order processing system that converts the unused buttons on customers' touch-tone telephones into a small computer terminal in order to place catalog orders.

The project will end a three-month Toronto test in mid-October. Approximately 2,000 regular catalog customers of Simpsons-Sears were offered the system, and, as of Tuesday [October 2, 1973], 453 customers had placed 1,130 successful orders.

No cost was given for the joint venture. Simpsons-Sears said that if initial test results prove favorable the system may be offered to customers across Canada.[30]

Of course, the Simpsons-Sears experiment did not involve two-way cable television. Apparently the firm offered regular catalog customers the opportunity to shop in this manner via direct mail, with return communication being via the unused buttons on the customers' Touch-Tone telephone. Yet experience gained from consumer response to the computerized catalog-order processing system would be helpful if Simpsons-Sears were to consider utilizing two-way cable as their communication link. Indeed, the material presented in the catalog could be converted into a presentation of the same information via display on the video screen. Viewers could then order by pressing appropriate buttons on the subscriber terminal (remote unit) in their home. Such an approach might reach not only Simpsons-Sears catalog buyers who are also CATV subscribers, but might attract the attention and interest of additional subscribers who do not customarily purchase through the use of a mail-order catalog.

If Simpsons-Sears found the computerized catalog-order system profitable and were to adapt it to two-way cable television, the implications would be most significant. Successful experience with this type of marketing in Canada might lead Sears Roebuck to adopt the approach in the United States where the influence on shopping behavior might be important indeed. Hence, the experiment merits close attention.

Call-A-Mart Supermarket, Louisville, Kentucky

In October 1973, *Newsweek* carried an article describing Call-A-Mart, Louisville's new and unique computerized supermarket. The firm was organized by Mark Weiss, a former employee of IBM. According to the report:

A Call-A-Mart customer simply consults a catalog that is issued by the store and updated weekly, selects what she wants, and phones in the order by number. ...The operator at Call-A-Mart punches up the numbers on a keyboard tied to an IBM computer. A computer-activated voice confirms the order and later that day it is delivered. ...While the computer is recording the order, it also prepares an invoice, keeps a running total on the tab and determines for the delivery-truck drivers the optimum route from warehouse to home. The computer is programmed so it knows all the addresses within Call-A-Mart's delivery zone, as well as the location of every traffic light and speed-limit sign. This permits it to lay out the most expeditious route from store to home.

The 8,500 customers Call-A-Mart has signed up pay $5 for the catalog. And while Weiss will not reveal his profit figures thus far, his estimate that Call-A-Mart customers average $35 to $40 in orders each week indicates that his firm may gross about $16.5 million this year....

The expenses of a computer, the automated warehouse from which he fills orders, and home deliveries are high, but Weiss is managing to stay competitive with such local chains as Kroger and A&P by resorting to a number of economies. He operates from just one location, saving the real estate and building costs involved in a chain operation. He keeps his inventory down to about 4,000 items vs. some 8,000 to 9,000 in a typical large supermarket. The overwhelming portion of sales in a given item are concentrated in one or two brands, so he whittled down the number of brands he carries—with one exception. "We have to carry a lot of breakfast foods," he says. "Kids are finicky." His operation also eliminates shoplifting, which amounts to about one percent of a normal supermarket's sales....

Newsweek notes that in organizing Call-A-Mart

Weiss gambled that a computerized supermarket would attract shoppers tired of the routine of driving, finding a parking space, bucking check-out lines and the rest of the hassle, and he has already signed up 8,500 customers in the four months since he started operating....

His customer list is growing rapidly. He reports,

In the last two weeks, we've expanded our customer list 25 percent and in the month prior to that we expanded 100 percent. ...The question is not whether we're going to expand, but where and when.[31]

Here, too, the system does not utilize two-way cable television but instead promotes the distribution of its catalogs in some unspecified way and relies upon telephone calls from the customer to the Call-A-Mart operator, who punches up the numbers on a keyboard tied to an IBM computer. Yet the system could utilize the two-way capability of cable in its operation, provided the Louisville CATV operator offered businessmen, in addition to two-way cable, a complete two-way system—including subscribers equipped with response terminals (remote units) and at the head end, a computer-controlled central data terminal.

Of course, if using two-way cable, Call-A-Mart would have to undertake promotion to get cable subscribers to buy the $5 catalog of merchandise, but there would appear to be no difficulty in accomplishing this task. Indeed, it might be possible to display the catalog pages on the video screen and thus either eliminate the need for the printed listing of merchandise or supplement the catalog with the television presentation of weekly specials to stimulate buying.

Experience gained by Mark Weiss may well provide a basis for the development of a system of electronic shopping for groceries and other household supplies. His innovative use of the computerized catalog is another significant step in the evolution of retailing.

Cable Catalog

In January 1974, Cable Catalog, a national experiment in consumer marketing via cable television, was announced by Formont Associates of New York City. Elizabeth Forsling Harris, chief executive officer of Formont, described the plan as follows:

> Cable Catalog will display merchandise from Neiman-Marcus in more than 50 communities with a total of 500,000 cable TV subscribers. The experiment is scheduled to extend through the first three calendar quarters of 1974.
>
> During the first stage of testing, Cable Catalog will offer subscribers a selection of 31 items from Neiman-Marcus. In the second stage, a second store and an additional 500,000 cable TV subscribers are to be added. And by the Christmas shopping quarter of 1974, Cable Catalog will be displaying merchandise from Neiman-Marcus and two other stores (yet to be named) to an audience of some 2 million subscribers.
>
> Subscribers will see a "motion picture catalog" of merchandise fed directly into the cable TV systems by means of video cassettes.

Elizabeth Harris noted that two-thirds of the cable TV market is located outside the nation's major marketing areas and that the service will not be offered in New York, Los Angeles, or San Francisco, adding that the experiment's success could mean a "major breakthrough in retailing techniques."[32]

Additional information about the Cable Catalog experiment from another source indicates that the merchandise featured on the 30-minute video tape may be ordered by mail. The tape cassette is being sent free to cable systems in 17 states reaching up to 500,000 cable subscribers. The 31 items of merchandise displayed on the tape range in price from a baby Neiman-Marcus print pillow retailing for $10 to a Kimberly three-piece pant suit tagged at $120. The tape, produced at a Neiman-Marcus store in Dallas and narrated by Chairman of the Board Stanley Marcus, was put together by Caravatt-Kleiman, Incorporated, New York, at a cost of $25,000. After the first nine weeks of the experiment, Ms.

Harris said that Formont will pay each system a nickel for each order received from a subscriber of that system, or 6 percent of the gross revenues taken in from people on that system who order merchandise, whichever is higher.[33]

Three aspects of this experiment are especially significant. The first is that merchandise normally sold in the Neiman-Marcus department store is being offered to cable subscribers. The second is that the products are promoted via video tape and made available to CATV system operators at no charge. It may be assumed that the advertising know-how of this leading department store is utilized in offering the 31 products to CATV subscribers. Third, the CATV operators stand to gain revenue up to 6 percent of the gross revenues generated by consumer purchases. If this method of selling proves to be attractive to CATV subscribers, therefore, the system operator will gain an additional—and potentially important—source of income.

The Cable Catalog experiment still lacks key elements of a two-way interactive system. While the merchandise is promoted via the cable video screen, the order is placed by mail rather than by pushing a button on a subscriber terminal. This arrangement is understandable, however, because as of January 1974 most CATV systems had not yet installed subscriber terminals and central data computer equipment, and the Cable Catalog approach nonetheless provides a way of gaining information on the responsiveness of consumers to products displayed and promoted on the cable video screen. Especially significant would be information on what types of product normally sold in a department store consumers would be willing to buy without examining the items firsthand and, if they fall in the category of clothing or footwear, trying them on.

Nothing is mentioned in the description of the system as to how payment is made—whether by check or via a Neiman-Marcus credit card. This aspect of the experiment is important; the tasks of providing capital to finance credit purchases and minimizing credit losses by sound administration are challenging problems that must be solved if profits are to be satisfactory. Further information about this experiment, therefore, should be of great interest to those considering the possibility of marketing via two-way cable television.

GiftAmerica

In June 1974 *Business Week* reported on a venture undertaken by Western Union to increase its unregulated business. In the fall of 1973, according to the report, Western Union began promoting

GiftAmerica, a service which offered consumers the convenience of dialing a toll-free number and having silver bowls, clock radios, and similar merchandise delivered anywhere in the United States within hours. Before undertaking this venture into retailing, Western Union hired Booz, Allen & Hamilton, management consultants, to survey the market. J. Walter Thompson Company was given the assignment of creating the advertising to promote Gift-America. Computers were programmed to handle orders, and more than 5,000 franchises were lined up as outlets.

When the first TV ads began running in the fall of 1973, according to an insider, the phones began to ring with calls from advertising men who said they could not understand the commercials. In June 1974, it was reported that J. Walter Thompson was no longer handling the GiftAmerica account and that the firm's first president, who had a technical communications background, had been replaced by a marketing executive.

The idea for GiftAmerica was developed by Russell W. McFall, Chairman-President, who submitted the proposal to three separate committees of Western Union officers before he found one that endorsed the idea. "Most of the officers felt that, with the scarcity of capital, the money could be better spent elsewhere," says the source. "Besides, nobody knew a damned thing about merchandising."

McFall still believes in GiftAmerica, but through March 1974 the operation had already cost $18.4 million. In its certification of the 1973 annual report, Price-Waterhouse noted in relation to GiftAmerica, "Recoverability of certain start-up costs is dependent upon the ability of a subsidiary to achieve and maintain a satisfactory level of operations."[34]

While GiftAmerica was apparently not promoted on cable television, the operation has some elements that relate to electronic marketing. The gifts were advertised on over-the-air television. Viewers could order gifts delivered to relatives or friends by dialing a toll-free number. The gifts were delivered within hours, presumably by franchisees. How payment was made is not indicated, although such a transaction could be handled by charging the purchase to one of several widely held credit cards.

GiftAmerica appears to be the type of service that might be marketed effectively via two-way interactive cable television. Television commercials would be cablecast to promote the use of GiftAmerica service by people wishing to send gifts to friends or relatives located in other parts of the country. Consumers could use alphanumeric subscriber terminals to specify the gift desired and to give the name and address of the person to receive it. This information could be transmitted to the local Western Union office and relayed to the franchisee servicing the community where the recipient lives. This dealer would then deliver the gift. The purchase again could be charged to widely held credit cards.

Before a firm like Western Union would be likely to consider the use of two-way cable for GiftAmerica, obviously, enough system operators would have to install subscriber terminals and central data computer equipment so that potential gift-givers could be reached via cable. CATV penetration of the 100 major markets would also have to be great enough to make cable a competitive promotional medium as compared with over-the-air broadcasting. Clearly these conditions are not likely to be achieved for some years. Yet the GiftAmerica concept appears to have considerable potential appeal to cable subscribers. Western Union's experience in stimulating consumer response with its present approach should provide interesting data on the feasibility of this idea.

CableMart, Incorporated

In August 1974, CableMart, Incorporated, had completed a period of four and one-half years of experimentation in the marketing of products to the subscribers of cable television systems.[35] CableMart (CM) had promoted these products through the use of direct mail, insertions in billing envelopes mailed to CATV subscribers, plus commercials presented via programs carried on cable systems. Items marketed in this manner included blenders, radios, digital clocks, floor polishers, and Kodak cameras, among others. Their average price was $40, although the firm had experimented with products selling for as much as $300.

CableMart, Incorporated, was a joint venture organized by Mark/James, Incorporated, a manufacturer's representative and syndicator, and Times-Mirror Communications, Incorporated, a gigantic publishing company with interests in several cable television systems. The function of CableMart was to experiment in cable television as an adjunct to the direct mail marketing of special products to the ultimate consumer.

In these tests the merchandise was offered via the "Cable Family Shopping Center," a division of CableMart set up to handle all promotion to individual subscribers. Cable television subscribers were provided with a membership card in this organization, which enabled them to charge their purchases of the merchandise that was offered. Orders were placed either by telephone or by mail since two-way interactive capability was not then available on the CATV systems over which CableMart was promoted. The products were delivered by a fulfillment house by mail or by parcel delivery. The customers sent their payment to CableMart, which handled the bookkeeping and the collection of accounts.

During a six-month period beginning April 1, 1971, CableMart conducted initial experiments in four California cities served by cable television systems which Times-Mirror Communications owned and which reached a total of 26,000 subscribers. The participating systems were located in Long Beach, Palos Verdes, San Clemente, and Escondido. The tests produced a small profit over operating expenses.

In the fall of 1972 the experiment was expanded to cover cable television systems in 18 states across the United States serving approximately 350,000 subscribers. This phase of the test included cable systems not owned by Times-Mirror Communications. This experience was successful enough to encourage officials of CableMart to plan further expansion of the Cable Family Shopping Center (CFSC) concept to cable systems owned by other Multiple Systems Operators (MSOs). More specifically, executives planned a step-by-step expansion beginning in February 1974 to include the systems owned by 12 additional MSOs in various parts of the United States with 1.5 million subscribers. The amount of time required to carry out this expansion plan was not specified, for much depended on the speed with which the Cable-Mart plan could be successfully introduced to additional cable systems, the availability of capital to cover developmental costs and to finance the credit purchases, and on other considerations as well.

In discussing the desirability of moving into full-scale operation with the Cable Family Shopping Center concept, executives considered a number of factors. The first was an anticipated increase of the cable television audience from approximately 9 million in August 1973 to between 35 and 40 million by 1980. Second, CM executives were aware of several field demonstrations of two-way interactive CATV under way in the summer of 1973. The demonstration by Theta-Com in El Segundo, California, was cited as an example. As CATV systems installed subscriber consoles making two-way interactive services available, electronic shopping would, of course, become even more convenient than the existing Cable-Mart plan, which involved the placement of orders either by telephone or by mail. The CableMart plan could easily be converted to utilize a two-way interactive system. Recognizing this trend, CableMart executives wanted to get enough experience in marketing over CATV to allow them to take full advantage of the two-way interactive capability when cable operators did begin to install push-button terminals for their subscribers. When such installations would begin on a significant number of CATV systems and how soon each system would make such equipment available to significant percentages of its subscribers was difficult to predict.

A third consideration in gauging the potential success of expanding the CableMart operation throughout the United States was the size of the audiences that could be reached by commercials promoting the Cable Family Shopping Center concept over those CATV systems that might be interested in signing up with the plan. Very few data were available on the size of the audiences reached by over-the-air programs carried on the cable to subscribers, however.

Rather than depending entirely on CATV operators to originate programs that would attract sizable audiences, CableMart executives were considering the development of some instructional programs of their own that might attract audiences which contained good prospects for CFSC products. A spokesman of CM explained this idea as follows:

> What we envision as a marketing tool would be, let us say, a full hour program presented at an appropriate time one day a week. For example, on one day present early in the afternoon a program on gardening, interior decorating, child care, personal care, as well as various types of cooking lessons. In the evenings present material that would be more attuned to the family interest or to men, such as golf lessons, bridge lessons, stamp collecting—any type of instruction that would have a broad appeal—and then at the end of the lesson, offer the viewers the opportunity of ordering every product that is used by the instructor on the program. In other words, if you present a bridge lesson, you could sell viewers the bridge table, the chairs, the cards, the instruction book on bridge, and so on. If we offer a series of 13 golf lessons by Arnold Palmer, we could then sell the viewers Arnold Palmer golf balls, clubs, bags, umbrellas....
>
> We believe that such programs will attract an interested audience for the Cable Family Shopping Center and that they will serve as an effective vehicle for selling goods via CATV. The cable operators are likely to favor such an approach since it promises to bring them in additional revenue.[36]

Financing and administering credit was another consideration of key importance to predicting how rapidly the Cable Family Shopping Center concept could be expanded into new markets. In the first test, CableMart had offered subscribers the opportunity of charging purchases to the credit card issued to members of the Cable Family Shopping Center. This policy required CableMart to finance credit purchases and administer the credit system. In the second test, therefore, CM executives decided to offer subscribers the option of charging their purchases either to Master Charge or to BankAmericard accounts. While this arrangement shifted the credit problem to the bank credit card firms, it was not particularly successful. It was discovered that many CATV subscribers did not have either of the two bank credit cards included in the CFSC plan. Moreover, reluctance to apply for such cards apparently had a depressing effect upon willingness to purchase.

The alternative of handling the credit problem through the Cable Family Shopping Center, as in the first experiment, required not only substantial capital as CableMart expanded to additional cable systems, but also know-how in handling credit business. Times-Mirror Communications, Incorporated, was not willing to contribute additional capital to help finance the expansion of the Cable Family Shopping Center concept. Accordingly, their interest in the CableMart joint venture was sold to Field Enterprises, Incorporated, of Chicago in November 1973. Mark/James, Incorporated, continued as a partner in the venture. Field Enterprises contributed not only the necessary capital, but also experience acquired from handling credit sales of encyclopedias by mail to ultimate consumers.

This change in ownership of CableMart delayed the expansion planned by executives of the firm. The step-by-step expansion program, which was to have begun in October 1973, actually was inaugurated in February 1974, and by August 1974 had been extended to three additional cable systems. This development underscores the importance of providing adequate capital and know-how in operating the credit aspects of any electronic shopping plan involving purchases charged to a credit card.

The CableMart operation is well worth watching. The firm is gaining experience in marketing via cable television. The past experience of its executives in direct mail selling provides useful background for the CM venture. While purchases are now being made by telephone or mail, a logical step would be for the firm to experiment with the use of two-way interactive cable television with one of the cable systems doing such developmental work. Since market tests of two-way services have been cancelled by CATV operators and equipment suppliers, as outlined above, such an opportunity may not present itself for several years. Nevertheless experience gained in such experimentation when the opportunity presents itself would pave the way for the conversion of the Cable Family Shopping Center plan to the use of two-way interactive cable, on a step-by-step basis, as cable operators equip their subscribers and their head-end facilities with the necessary hardware.

Cable Response Network, Incorporated[37]

In November 1973 Shop-at-Home service was inaugurated on 40 cable systems with approximately 500,000 subscribers exposed to the program. Shop-at-Home is the title of a 30-minute video tape produced by a new company called Cable Response Network, Incorporated, of New York. It was scheduled to run from November 1 through the Christmas holidays. This show was the first of a series of video tapes that will promote various types of merchandise. The CATV system was provided with the show free, with the agreement that it would be run at least once a day over the operator's origination channel.

The Shop-at-Home program that was presented November 1 through December 1973 described, in its thirty minutes, a dozen different products that were presented in an entertaining manner by a host and hostess, singer Johnny Andrews and actress Suzanne Astor. Each item was followed by a slide listing its number and price. The viewer could order by phoning a toll free number or by writing to the Shop-at-Home address. Master Charge and Bank-Americard were accepted. The program was produced at Windsor Total Video, New York City.

The products promoted on the program were especially selected by Mail Mart Associates, a mail order house, for sale during the Christmas season. The following items were included at the prices shown:

Heartlace necklace	$10.95	Supa-sharp knife	$12.70
3-player chess	16.70	Home and apartment	
Lazy pad	52.95	tool kit	16.95
Wall of shelves	67.45	Touchlite	32.45
Supermatch-acrylic	11.45	Slimaster cycle	20.95
Supermatch-crystal	26.95	Flight bag	16.45
5-year flashlight	7.95	Alarm strong box	41.45

Shipping charges were included in the prices, and customers were told to allow three to four weeks for delivery. Mail Mart filled and shipped the orders.

According to the plan, the cable system received 15 percent of the sale in return for running the show. A free gift was also provided for the cable system to give to a lucky CATV viewer as part of a plan to stimulate interest in Shop-at-Home.

Cable Response Network (CRN) would not disclose its gross margin percentage, but the profit potential might be limited since CRN bears all the costs of program production, duplication, and distribution.

Principals in CRN include Robert W. Rawson, who has a background in the mail order business, and Robert Schultz, who heads a CATV audience-rating service known as Video Probe Index.

In February 1974 Robert Schultz, vice-president, Cable Response Network, Incorporated, gave the following assessment of the Shop-at-Home experiment:

> Almost 40 systems ran the show based on a verbal description given them on the telephone. The tape was then sent them with the option of running or returning. Only a few saw the tape and then decided not to run the show. The subscriber potential exposure was estimated at 500,000. The viewership, I would imagine, was quite small even though a number of systems ran it more than the required once-daily.
>
> I might add that the results were disappointing. Whether this was because of the audience size, items selected, lack of promotion, or just plain resistance during a period of economic uncertainty, is hard to say.
>
> What I can say is that we will not invest again in a general merchandise show. In the future we will concentrate on a specific interest-group show.[38]

It would appear that a program such as Shop-at-Home would need to be promoted to cable subscribers in order to get them to watch the cable system's program origination channel. CRN did not do this; instead the firm relied upon the cable operator to publicize the program. Obviously the cable systems did not deliver a large enough audience to permit an adequate test of the Shop-at-Home concept. The plan of appealing to special interest groups in future programs also appears to be preferable to relying upon a general merchandise show.

Telephone Computing Service, Incorporated

A new home computer service using Touch-Tone telephones as the input medium and a voice-response system for output was placed in operation in Seattle on June 4, 1973.

The service, called *In-Touch*, was provided by a new company, Telephone Computing Service, Incorporated, a subsidiary of Seattle-First National Bank. Home subscribers paid $6.50 a month for a variety of data-processing services, including automatic payment of bills by telephone, income tax preparation, and operation as a four-function calculator.

For the monthly fee subscribers received 100 minutes of usage and paid four cents a minute for more. The system used all twelve buttons on the Touch-Tone telephone. Templates over the buttons instructed subscribers in the use of the system. Each user had a personal code number that activated the phone-computer link.

There were six major services: funds transfer for bill payment, family budgeting (data acquisition by phone and a weekly mailed printout), income tax processing, calendar reminder service, household record-keeping, and calculator service. As a calculator, the system accepts Touch-Tone input and supplies results by voice answer-back.

After four months' experience with this service, Seattle-First National claimed that response had been "fantastic" with "several hundred" subscribers signing up since the service was introduced in June 1973.[39] The one problem then evident with the service, according to Seattle-First National, was convincing people that the telephone-linked equipment can keep reliable records.

After six months of operation, however, the president of Telephone Computing Service, Incorporated, announced that the firm would go out of business on December 29, 1973. It was explained that a major obstacle was the requirement that customers have Touch-Tone telephones. While company executives believed the procedures used to operate the computer via the Touch-Tone telephone were simple, they proved to be too complicated for many people. Other inconveniences included lack of weekend computer time, too small a pool of merchants to whom bills could be paid using the system, and too high a service charge, among others. The most popular aspect of the service was its ability to pay bills.[40]

In spite of the failure of this venture, In-Touch is an example of a type of service that might be modified, simplified, and offered over a two-way cable system. Since the payment of bills was popular with customers, the service might stress this function initially. As other services customers want are identified, they, too, might be added. The growing use of electronic computers by students might hasten acceptance of this function as time passes. The possibility of a joint venture involving a CATV system with two-way capability and a bank like Seattle-First National might be worth considering.

If the CATV system installs two-way subscriber terminals, they would serve as well as, or better than, the Touch-Tone telephone as the communication link with the computer. A CATV system might offer pay TV as well as other services that—combined with home computer service—might build a package appealing to a large enough percentage of the cable subscribers to make the venture profitable. Here is a long-range possibility that is worth keeping in mind.

Electronic funds transfer systems (EFTS)

In September 1973 *Business Week* published a special issue dealing with "new banking," which suggests a revolutionary change in the methods by which people and business firms handle their money transactions. This system is popularly referred to as the cashless-checkless system. More specifically, it might be described as paperless electronic funds transfer systems (or EFTS). According to *Business Week,* Electronic Funds Transfer Systems will be banking's next big step.

> Evolving rapidly, the basic technology and system design for an electronic funds transfer system (EFTS) is ready to go. Major questions of ownership, control and regulation remain to be settled. But both bankers and data-equipment suppliers agree that any system must include at least the basic processing elements in the diagram: (1) credit cards, (2) point of sale terminals, (3) cash dispensers and remote tellers, (4) communications link (initially dial-up or leased telephone lines, later cable TV or specialized communications carriers), (5) local switches (local message-switching computer centers), (6) bank computer centers, (7) verification centers (for a fast initial check of card validity, credit standing, transaction limit, and frequency-of-use pattern), (8) automated regional clearing houses, and (9) national data network.[41]

As an example of the potentialities of EFTS *Business Week* reports that in Columbus, Ohio, in September 1973, after a nine-month test in suburban Upper Arlington, City National Bank and Trust Company was forging ahead with plans to install 125 terminals in 60 major stores and supermarkets in the Columbus area. This arrangement permits "paperless" payments by customers with credit cards.

Also it was reported that automated clearing houses in Atlanta and San Francisco were distributing company payrolls electronically and automatically crediting employees' bank accounts. Banks were urging customers to use single multi-payment checks to pay many bills at once. The Atlanta Payments Project had further plans to set up a regional electronic payments system that would extend electronic cashless, checkless transactions to the retail sales level.

Again, in Pittsfield, Massachusetts, seven financial institutions, including a federal credit union, savings bank, and commercial banks, were cooperating on a joint study to set up a city-wide electronic payment system. If these plans go through, Pittsfield could become a model for electronic funds-transfer systems in smaller towns.

While in the beginning the communications link in EFTS systems will be over ordinary dial-up or leased telephone lines, cable television systems could compete effectively for this business at such time as they are equipped to handle two-way interactive television service. Also, the availability of an electronic funds transfer system in the area served by a cable operator would greatly facilitate the handling of payments from subscribers utilizing electronic shopping, premium entertainment, or other two-way cable services.

Although there are, of course, serious problems to be overcome in developing the EFTS, a number of bankers close to the advance work in the field believe that paperless EFTS systems are coming on so fast that they are shortening their own timetables. For example, John F. Fisher, Vice-President of City National Bank in Columbus, states, "We believe that by 1977 five percent of the metropolitan areas of the country will have EFTS in place and by 1983 that will grow to 35 percent of the metropolitan areas."[42]

Fisher was talking about the systems that put terminals at the point of sale, directly triggering a process to electronically debit the customer's account and credit the merchant's account without generating any paper records except as a receipt. With such systems, merchants could also issue cash and post credits to a customer's account for returned merchandise.

Potential Market Demand for Two-Way Interactive Services by 1990[43]

The foregoing discussion of demonstrations and experiments with two-way interactive cable television makes it clear that this innovation is still in the early stage of development and testing. As of August 1975 two-way services were not being offered on a commercial basis.

Nevertheless interactive services are expected to become an increasingly important factor in the growth of cable television in the decade ahead. Accordingly, it is interesting to speculate on their potential. Of special significance is a 1973 study by Paul Baran and Andrew J. Lipinski of Cabledata Associates, Incorporated, Palo Alto, California, which explores the dimensions of a projected new industry involved in providing information services in the home. Paul Baran describes this study and its findings as follows:[44]

As a part of this study, we considered entirely new demands for communications of a type presently not provided. We obtained the cooperation of a group of experts whom we felt would be the people most likely to provide a useful forecast of developments in two-way information services. The experts were individuals who generally were proficient in the technological aspects of the future systems and were often broadgauged "company philosopher" types. Since there are no real experts on new communication services, our respondents were people who are practical technologists, yet also imaginative and not afraid of making estimates for currently nonexistent services.

To obtain the views of these experts, we used the Delphi technique, which provides for the use of questionnaires to collect judgmental data from expert respondents who remain anonymous with respect to one another.

The panel's judgments were elicited by means of two questionnaires. In the first questionnaire, respondents were given a list of 30 potential new home information services, and were asked to estimate such things as the most likely year of mass introduction to the United States and the percentage penetration of households five years later.

In the second round, respondents were given the results of the first-round questionnaires and were asked to re-estimate all numerical forecasts in the light of the other panelists' responses.

The results indicated that *the biggest market in 1989 in terms of dollar value will be for plays and movies from a video library. This service was forecast to have a dollar value of $2,829,000,000.*

The next biggest market would be for computer-aided school instruction ($2,047,000,000).

Other services with markets estimated to be over the $1 billion mark were cashless society transactions, person-to-person communications involved in paid work at home, computer tutor systems, and adult evening courses on television.

The total value of the market for all 30 services would be about $20 billion. Some of the services were overlapping. For instance, electronic shopping at home overlaps cashless-society transaction, but counterbalancing this "double accounting" is the realization that the list of 30 services is incomplete. Omitted were such major services as home security systems and utility meter reading. The various services were as follows: (value estimates in millions of dollars)

Education: The category of services for the home that will produce the most revenue for cable TV is expected to be educational services. The four separate educational services included in this category are: computer-aided school instruction, $2,047; computer tutor, $1,414; correspondence school, $943; and adult evening courses on TV, $1,131. These four services differ chiefly in the amount of interaction they offer between the student and the system.... (Total value $5,535).

Business conducted from the home: Services enabling people to conduct business from the home are expected to account for 20–25% of total anticipated revenues for these broadband services. Five separate services are included: person-to-person paid work at home, $1,713; secretarial assistance, $707; access to company files, $46; computer-assisted meetings, $707; and banking services, $566.... (Total value $3,739.)

General information access: The next major category of services provides "access to information" and includes nine services: dedicated newspaper [a newspaper, the organization of which has been predetermined by the user to suit his preferences], $849; fares and ticket reservations, $124; daily calendar and reminder about appointments,

$292; newspaper, electronic, general, $200; legal information, $285; weather bureau, $228; bus, train, and air scheduling, $79; library access, $95; index of all services served by home terminal, $106.... (Total value $2,258.)

Shopping facilitation: The next major category concerns ways to make shopping easier. This includes five separate services: cashless-society transactions, $1,810; shopping transactions (store catalogs), $584; grocery price lists, information, and ordering, $566; special sales information, $354; and consumers' advisory service, $354.... (Total value $3,668.)

Entertainment: Three separate services are included under the general heading of entertaintment. These include: plays and movies from the home library, $2,849; past and forthcoming events, $130; and restaurants, $35.... (Total value $3,014.)

Person-to-person communications: While we have good person-to-person communications systems today, there are a number of primarily communication activities that might be done better by new communications systems than by present means. These include new services for message recording, $106; household mail and messages, $707; mass mail and direct advertising mail, $0; and answering services, $743.... (Total value $1,556.)

Our panel of experts forecast that all these services would likely be introduced between 1975 and 1990, with the most likely date for introduction of these services being about 1980.[45]

A summary showing 30 information services for the home with forecasts as to when they may be introduced and what they may be worth in dollars is presented in Table 7.

Summary and Conclusions

Offering two-way interactive cable services has been suggested as a possible solution to the problem of the cable operator attempting to penetrate major urban markets, where three network signals, as many as three independent stations, plus an educational station are all available over the air via rabbit ears. If the operator is to build penetration of such television viewing markets, clearly he needs to offer some services that will develop a competitive edge for him. The following paragraphs summarize the promise of two-way cable services and the problems involved in developing them.

1. An extensive list of possible two-way services has been suggested for consideration by cable operators, whose problem is to determine which are likely to have greatest appeal in building penetration in major markets and in adding to system revenues. Estimates of the cost of providing such services are essential in reaching a decision as to the sequence in which potentially promising services might be introduced.

Table 7

FORECAST FOR INTRODUCTION OF INFORMATION
SERVICES FOR THE HOME

Service	Estimated Year of Introduction			Projected Value of Market in 1989 (in $ Millions)
	Early Estimate	Middle Estimate	Late Estimate	
1. Plays and movies from a video library	1975	1980	1985	2829
2. Computer-aided school instr.	1975	1982	1987	2047
3. Cashless-society transactions	1975	1980	1990	1810
4. Person-to-person (paid work at home)	1980	1985	1990	1713
5. Computer tutor	1975	1980	1990	1414
6. Adult evening courses on TV	1975	1980	1985	1131
7. Correspondence school	1978	1984	1990	943
8. Dedicated newspaper	1980	1983	1990	849
9. Answering services	1975	1980	1985	743
10. Computer-assisted meetings	1975	1980	1985	707
11. Household mail and messages	1980	1985	1990	707
12. Secretarial assistance	1975	1980	1985	707
13. Shopping transactions (store catalogs)	1977	1985	1990	584
14. Banking services	1975	1980	1985	566
15. Grocery price list, etc.	1975	1980	1990	566
16. Special sales information	1975	1982	1990	354
17. Consumers' advisory service	1975	1980	1985	354
18. Daily calendar, etc.	1980	1983	1985	292
19. Legal information	1980	1985	1990	285
20. Weather Bureau	1975	1980	1980	228
21. Newspaper, electronic, gen.	1980	1985	1990	200
22. Past and forthcoming events	1975	1982	1990	130
23. Fares and ticket reservation	1975	1980	1985	124
24. Message recording	1975	1980	1985	106
25. Index, all served by home territory	1975	1980	1985	106
26. Library access	1980	1985	1990	95
27. Bus, train, air scheduling	1975	1977	1980	79
28. Access to company files	1980	1985	1990	46
29. Restaurants	1975	1980	1985	35
30. Mass mail/direct adv. mail	1980	1990	1995	0

Source: Paul Baran in "30 Services that Two-Way Television Can Provide," *The Video-Cassette and CATV Newsletter*, Jan. 1974, p. 10.

2. Subscriber response services, perhaps with shared voice return channels, seem likely candidates for home use in the next five years. The investment cost for the basic two-way equipment required would amount to roughly $150 to $340 per subscriber, over and above the $125 per subscriber calculated for conventional one-way cable service. Two-way services that could be provided by this equipment would include remote shopping, interactive entertainment and instructional programming, selection of subscription or limited-access channels, and audience counting for advertisers and programmers.

3. With the equipment mentioned above and with operating costs based on reasonable assumptions, a cable operator would need additional monthly revenues of between $4.50 and $13.00 per subscriber to break even on two-way response services. This means doubling or tripling his present monthly revenue from one-way television distribution. Most of the added revenue would have to come from increased monthly subscriber fees, although advertisers, business firms, utilities, schools, and government users would pay for services of benefit to them. Providing a mix of response services supported by home subscribers and business and government users appears to be a better strategy for the cable operator than supplying a single service alone. Expected revenues from specific services cannot be estimated at this time, since no real field experience or evidence of consumer demand is yet available.

4. A number of demonstration projects have been undertaken or are in progress with the purpose of uncovering technical problems in providing two-way services and getting better information on anticipated costs. We have reviewed Mitre Corporation's test in Reston, Virginia, as well as the plans for the operational experiment in Stockton, California; the Jonathan/Chaska Community Information System Project; the El Segundo Pilot Project; the Orlando, Florida, test; and the Irving, Texas, experiment. Unfortunately, plans for market tests of consumer demand for two-way services have been delayed or cancelled because of the impact of unfavorable economic conditions on the cable industry. When they will be executed or reinstated is unknown.

Although several experiments supported by government agencies and involving the delivery of various social services are going forward, these are not expected to lead to commercially viable uses of two-way.

Installation and management of two-way systems in housing developments appears to offer a promising opportunity for equipment producers to gain experience in the operation of interactive systems and to accumulate information on consumer demand for intrusion, fire, medical emergency, as well as other services.

5. Several firms have taken initial steps that might be regarded as preparing the way for eventual development of two-way electronic shopping. Elements included in their approach are computerized catalog ordering systems, promotion of purchases by direct mail, over-the-air television, or cablecasting; placement of orders by Touch-Tone telephone, by mail, or by toll-free telephone; payment by charge to credit card or C.O.D.; delivery by fulfillment house, company-owned and franchise-owned trucks, or by parcel post. None provides for placement of orders via two-way cable subscriber terminal, possibly because cable systems have not yet installed such equipment in a significant percentage of subscribers' homes.

6. Various combinations of these elements of two-way electronic shopping, which may be thought of as evolutionary steps toward fully developed electronic shopping services, have been included in experiments by the following firms:

Simpsons-Sears and IBM of Canada. Three month test of computerized catalog-order processing system using Touch-Tone telephone for placement of customer orders. Of 2,000 regular catalog customers, 453 placed 1,130 successful orders. No costs reported.

Call-A-Mart Supermarket of Louisville. Interim report on first year's operation of unique computerized supermarket where consumers phone in orders for grocery products using $5 catalog for information on items and prices. Had signed up 8,500 customers; estimated gross profit about $16.5 million for the year.

Cable Catalog. Experiment with motion picture presentation of 31 items of merchandise from Niemann-Marcus during 1974. Thirty-minute videotape sent free to cable systems in 17 states reaching 500,000 cable subscribers. Orders placed by mail. No report on sales and costs of experiment.

GiftAmerica. A venture by Western Union offering consumers the convenience of dialing a toll-free number and having gift merchandise delivered anywhere in the United States within hours. Promoted via over-the-air television commercials. After six months it was reported the venture had cost $18.4 million. No report of revenues generated by the promotion.

Cable Response Network, Incorporated. Four months' experience with Shop-at-Home service offered on 40 cable systems with 500,000 subscribers reported upon in Feb. 1974. Thirty-minute Shop-at-Home program was supplied to cable operators via video tape. It promoted 13 different products believed to be potential gift merchandise for Christmas shoppers. Orders placed by toll-free number or by mail. Purchases charged to credit card; fulfillment by Mail Mart direct mail order house. Results were disappointing. Possible reasons: small audience for program, items offered, lack of promotion of Shop-at-Home program to stimulate subscribers to view program. Changes in approach planned.

CableMart, Incorporated. Report on 4½ years of experience in marketing products by cable television via the Cable Family Shopping Center. Items promoted chiefly were hard goods, priced $40 to $300. Promoted by mail and via cable television commercials. Orders were placed via phone or mail; purchase was charged to credit card; delivery by established fulfillment houses. After 4½ years CFSC was being promoted over cable television systems in 18 states with approximately 350,000 subscribers. Results: a spokesman claims the venture had broken even up to that point. Plans made for expansion to 12 additional MSOs in various parts of United States with 1.5 million subscribers when market conditions are favorable.

7. In building a package of two-way cable services, a cable operator might combine some type of electronic shopping plan with a home computer service and a service providing checkless-cashless banking transactions. There have been two interesting experiments in this kind of service.

Telephone Computing Service, Incorporated. A new home computer service using Touch-Tone telephones as the input medium and a voice-response system for output was placed in operation in Seattle in June 1974. Home subscribers pay $6.50 per month for data-processing services, including automatic payment of bills by telephone, income tax preparation, and operation of a four-function calculator. Although initial response was favorable, the service was withdrawn after six months. If changes were made to correct problems encountered in the original operation, however, this service might well be considered by a CATV operator for inclusion in a two-way package offer to subscribers.

Electronic Funds Transfer System (EFTS). A number of experiments with "cashless-checkless" banking were underway in September 1973. Revenue and cost information were not available.

(a) After a 9-month test, a bank planned to install 125 terminals in 60 major stores and supermarkets in Columbus, Ohio. This development would permit so-called paperless payments by customers with credit cards.

(b) Seven financial institutions in Pittsfield, Massachusetts, were cooperating in joint study to set up city-wide electronic payment system.

(c) In Atlanta and San Francisco automated clearing houses were distributing company payrolls electronically and automatically crediting employees' bank accounts. Banks were urging customers to use single multipayment checks to pay all bills at one time.

8. Because review of these experiments with various types of two-way service does not provide any significant information as to the revenues and costs that operators may anticipate in providing such facilities, prospective profits cannot be estimated at this time. Certain of the tests may provide such essential data in the future,

however, and cable operators would be well advised to watch closely for publication of the results of these interesting and significant experiments.

9. Although two-way interactive cable services were not being offered commercially as of June 1975, observers expect them to become increasingly important factors in stimulating the growth of cable television penetration in the decade ahead. Hence it is important to review the predictions of experts as to the timing of the introduction of various two-way services and the anticipated growth in the demand for these services by 1989. In a study using the Delphi technique such experts made forecasts of the potential market for 30 possible two-way services by 1989.

The five individual services showing the most favorable prospects (in order of projected value of their markets) are the following:

Service	Projected Dollar Value of the Market in 1989 (in Millions)
Plays and movies from video library	$2,829
Computer-aided school instruction	2,047
Cashless society transactions	1,810
Person-to-person (paid work at home)	1,713
Computer tutor	1,414

Shopping transactions were estimated at $859 million, which placed them thirteenth on the list. Grocery price lists, which bring in revenues of $566 million, were fifteenth. Special sales information was estimated at $354 million (sixteenth), the same as providing a consumers' advisory service.

10. When the individual services were classified as to type, their projected potential markets ranked as follows:

Type of Service	Projected Dollar Value of the Market in 1989 (in Millions)
Educational services	$5,535
Business conducted in the home	3,739
Shopping facilitation	3,668
Entertainment	3,014
General information access	2,258
Person-to-person communications	1,556

11. These projections were made before the recession of 1974–75, which has served to delay the development of two-way interactive television, and therefore they are probably optimistic. The rank order of potential markets by size would nevertheless still appear to be significant, although it will probably take longer for these markets to develop to the magnitudes estimated.

1. *Credit Cards*
     Like credit cards in use today, but for EFTS use they will include some positive identification check, i.e., photograph or confidential code memorized by the bearer.

2. *Point of Sale Terminals*
     Ranging from an ordinary telephone to an electronic cash register using a mini-computer.

3. *Cash Dispenser and Remote Tellers*
     The simplest versions simply dispense a fixed amount of cash. More complex terminals will also record deposits and transfer funds between accounts.

4. *Communications Link*
     In the beginning, communications will be over telephone lines. Later specialized carriers or cable TV companies could compete.

5. *Local Switches*
     Local message-switching computer centers will be able to handle thousands of simultaneous transactions almost instantaneously. Such centers will relay transactions to the appropriate financial institution of credit validation system.

7. *Verification Centers*
     POS terminals will have rapid access to files for a fast check of card validity, credit standing, credit limit, and frequency-of-use pattern.

6. *Bank Computer Centers*
     Banks and thrift institutions participating in EFTS will have access to an on-line 24-hour computer system capable of verifying and processing transaction messages, debiting customers' accounts, storing transaction data for billing, and issuing credits to vendors' accounts.

8. *Automated Regional Clearing Houses*
     The ACHs will switch all interinstitutional transactions involving funds transfers. They also might be used to switch interregional transaction inquiries and data through a national financial network such as the Federal Reserve.

9. *National Data Network*
     High-speed, high-accuracy financial data networks which interconnect the automated clearing houses and credit verification systems.

Fig. 4. Basic processing elements of the Electronic Funds Transfer System (EFTS). (Source: "The Quickened Pace of Electronic Banking," *Business Week*, Sept. 15, 1973, p. 117.)

NOTES

1. The technical aspects of this discussion are adapted from Walter S. Baer, *Interactive Television: Prospects for Two-Way Services on Cable* (Santa Monica, Calif.: Rand Corp., Nov. 1971), Sections I, II, and III.
2. FCC, *Cable Television Report and Order*, Sec. 129, 37 Fed. Reg. 3252, 1972.
3. Carl Pilnick and Walter S. Baer, *Cable Television: A Guide to the Technology* (Santa Monica, Calif.: The Rand Corporation, 1973), pp. 39–41.
4. Baer, *Interactive Television*, pp. 8, 23–24.
5. A system's communication capacity is measured by its bandwidth in cycles per second (or in the more modern units of "hertz," abbreviated Hz). Each U.S. standard television channel requires a large frequency bandwidth of 6,000,000 hertz, usually stated as 6 Megahertz, abbreviated to 6mHz. Thus, the FCC's 20-channel requirement actually means a usable bandwidth of 20 x 6, or 120 mHz. (Pilnick and Baer, *Cable Television...*, p. 13.)

   Some services, like simple opinion polling, demand only a single bit (a bit—binary digit—is the common unit of information)—a yes or no response. Others, like remote shopping, may require that a few alphanumeric characters, or several tens of bits, pass upstream from subscriber to headend. Such data or message services may require only about 100 hertz (Hz) per subscriber upstream. In contrast, a voice channel requires 3–4 kilohertz (kHz), and standard color video transmission uses 6 mHz. Thus color video origination by one subscriber might require more upstream bandwidth than returning digital data from 50,000 households.

   Baer uses the term "narrowband" to describe services requiring transmission bandwidths less than 3 kHz; "voiceband" for 3 to 4 kHz; "wideband" for 4 to 1,000 kHz; and "broadband" for bandwidths above one megahertz.
6. Baer, *Interactive Television*, pp. 14–16.
7. Pilnick and Baer, *Cable Television...*, pp. 37, 39.
8. Baer, pp. vi, vii.
9. *TV Communications*, June 1972.
10. Information derived from a videotape of the Mitre Corporation demonstration, plus news report in *Broadcasting*, August 27, 1973, p. 47.
11. Baer (*Interactive Television*) explains that "frame grabbing," "frame snatching," or "frame stopping" is a technique that involves transmitting a single frame downstream for recording and display as a still picture on the television screen. It permits displays of pictures, charts, and drawings, as well as alphanumeric characters.

    Functionally, a frame-stopping device contains the same sort of receiver and address decoder as a subscriber data terminal, plus electronic or magnetic storage to record a single frame, and a control unit for recording and displaying the frame on a standard television receiver. None of this is technically difficult, according to Baer, but frame storage is relatively expensive.

    The frame-stopping device shown by Mitre in Reston was the first to be publicly demonstrated. It was developed just as a terminal for computer-aided instruction and employs an inexpensive helical scan videotape recorder for frame storage. A frame addressed to the subscriber is recorded at his terminal. It then is continuously replayed from the videotape recorder onto his television receiver until a new frame is ordered.

    Similar frame-stopping terminals might cost about $1,500 commercially at the present time, although Mitre expects the development of low-cost videotape recorders to reduce this figure dramatically in two to five years.

    In the Mitre system, Baer notes, one individually addressed frame can be transmitted every 1/60th of a second on a single video channel. Thus, if the subscriber retains a frame for 10 seconds on the average, one 6mHz downstream channel could serve 600 subscribers simultaneously. (*Interactive Television*, pp. 29, 30.)

In a letter dated May 28, 1975, however, a company spokesman reports that the videotape recorder (VTR) described above is no longer considered to be a likely frame-stopping device. Mitre's new work is based on the use of centrally located (rather than home-located) frame storage, using large scale integration (L.S.I.) shift-register storage (i.e., digital storage technique using large scale integrated circuits).

12. This section is based upon a talk by Timothy Eller, Mitre Corporation, Feb. 1, 1974, before the Faculty Seminar on Telecommunications, at the University of Michigan, and upon news articles in *Broadcasting*, Aug. 27, 1973, p. 47, and Feb. 25, 1974, p. 53.

13. "Mitre Selects System in California for Two-Way Test," *Broadcasting*, Feb. 25, 1974, p. 53.

14. This updated information is based upon correspondence with Timothy Eller, Group Leader, Computer Systems, Mitre Corporation, May 28, 1975, supplemented by a telephone interview on June 6, 1975.

15. This section is based upon information gathered during a personal interview with E. D. McCormick, President, Community Information Systems, Inc., August 1973, and on brochures and other information supplied by CIS. Important information was also drawn from the report, *The Jonathan-Chaska Community Information Systems Experiments* prepared for the U.S. Department of Housing and Urban Development, by Community Information Service, Inc., June 1973.

16. CIS, Inc., *The Jonathan-Chaska Community Information System Experiments*, June 1973, p. I-17–I-23.

17. *Ibid.*, pp. iv-vi, I-9.

18. This section is based upon a visit to the demonstration at the Hacienda Motel, El Segundo, in June 1973, together with reports of the project by Theta-Com officials and brochures describing the Interactive Cable TV Project.

19. Robert W. Behringer, "Blue Sky to Cash Flow: Market Study," reprint of a paper presented at the NCTA Convention, Chicago, Illinois, May 15, 1975, pp. 5–8. Richard T. Callais, "Subscriber Response System: Progress Report," *Subscriber Response System*, brochure, Theta-Com of California, June 1973, pp. 10, 12.

20. *Video Publisher*, January 28, 1975, p. 7.

21. Telephone conversations with H. R. Goodman, Manager, Multiplex System Program, Micro-Electronic Products Division, Hughes Aircraft Company, May 29, 1975.

22. Based on a telephone interview with Ed Harmon, Electronic Industrial Engineering, Inc., North Hollywood, California, May 24, 1973.

23. Based upon a telephone conversation with Marshall Savage, Marketing Service Manager, Electronic Industrial Engineering Inc., North Hollywood, California, on May 29, 1975.

24. This report is based upon a telephone conversation with Jim Smith, National Sales Manager, CATV Division, Tocom, Inc., May 24, 1973. He supplied a brochure describing the Tocom system and a reprint of an article in the *Dallas Times Herald* business section by Bronson Harvard, entitled, "Irving Will Experiment with Big 'Little Box'," Nov. 19, 1972, pp. 1–2. Adapted courtesy the *Dallas Times Herald*.

25. Based on a telephone conversation with Jim Smith, National Sales Manager, CATV Division, Tocom, Inc., May 30, 1975.

26. Adapted from *Cable Television: Takeoff into Sustained Growth* (New York: Samson Science Corp., a subsidiary of Quantum Science Corp., 1972), p. 14.

27. *CATV Newsweekly*, Dec. 9, 1974, p. 32.

28. *Videocassette & CATV Newsletter*, June 1974, p. 11.

29. Letter from Timothy Eller, Mitre Corp., May 28, 1975.

30. "Sears Puts Wish Book Computer in Home," *Detroit Free Press*, Oct. 4, 1973.

31. "Supermarkets: Dialing for Doughnuts," *Newsweek*, Oct. 22, 1973.

32. "Cable Catalog to Test Selling via Cable TV," *Merchandising Week*, Jan. 28, 1974, p. 19.

33. "Shop by Cable," *Broadcasting*, Feb. 4, 1974, p. 43.

34. "GiftAmerica," *Business Week*, June 8, 1974, pp. 72–74.

35. Summarized from a case history prepared by the author entitled, *CableMart, Inc.*, Marketing Cases Series No. 107, Graduate School of Business Administration, The University of Michigan, Aug. 1974.

36. Scott, *CableMart, Inc.*, p. 19.

37. Adapted from *Video Publisher* 3, No. 12 (1973):1; and *Broadcasting*, Nov. 12, 1973, p. 47.

38. Letter to the author, February 13, 1974.

39. *Wall Street Journal*, Oct. 4, 1973, p. 1.

40. *New York Times*, December 29, 1973, p. 31.

41. "The Quickened Pace of Electronic Banking," *Business Week*, Sept. 15, 1973, pp. 116 ff.

42. *Ibid.*

43. This section is adapted from "30 Services that Two-Way Television Can Provide," *The Videocassette & CATV Newsletter*, Box 5254, Beverly Hills, CA 90210, pp. 1–10.

44. Paul Baran is president of Cabledata Associates, Inc., 701 Welch Road, Palo Alto, CA 94304. He is the author of a definitive report on two-way television, "Potential Market Demand for Two-Way Information Services to the Home: 1970 to 1990." His work in the area of two-way television is an outgrowth of an earlier study on the future of the telephone system made for American Telephone and Telegraph Company.

45. "30 Services that Two-Way Television Can Provide."

III

ANALYSIS OF ALTERNATIVE
PAY-CABLE SERVICES

In reporting on the demand for two-way cable television in Jonathan/Chaska, Community Information Services, Incorporated, noted that premium programs (such as feature movies, musicals, sports events) were among the three extra services most desired by prospective subscribers. This finding is supported by a Stanford Research Institute report which concludes that pay-TV will increase program diversity by providing limited interest programs to those who are willing to pay, while not denying others the benefits of "free" television.

The Stanford report predicts that 30 percent of all TV households will be pay subscribers by 1985: There will be 1.5 million pay-TV households by 1976, and a million more in each succeeding year, surpassing 25 million by 1985. As a result, the report says, revenues during the mid-1980s will approach $4 billion annually, up from $200 million in 1976.[1]

According to SRI, economic analysis of pay-cable TV systems predicts there will be an after-tax return of more than 20 percent. Their study suggests that the relatively high profitability of pay cable provides a strong incentive for investors and predicts that capital will be available to promote the rapid expansion of this service.

According to SRI, CATV should dominate pay television within ten years, because its multichannel capability permits frequent offering of a wide variety of programs. In the major markets,

---

NOTE: This section is based upon research done by Peter E. Robinson, Research Assistant, under the direction of the author. Information was collected by personal interview as well as from published sources. The basic report was prepared as of March 1974.

however, CATV will encounter strong competition from over-the-air pay television operations, because they can be installed more rapidly and will thus dominate the near-term.

Accordingly, the cable television system operator who is interested in increasing subscriber revenues should carefully consider the types of pay-cable services available and the services or benefits each offers to his CATV system.

Historical perspective

The cable television industry hopes incremental revenue from pay-cable will substantially increase profit margins on existing cable systems as well as increasing the rates of return on capital investment, thus providing incentive to wire the major metropolitan areas.

For an infant industry this is a formidable task. The industry only came to life during the last week of January 1973, according to the *Pay TV Newsletter*.[2] During that week three major pay-TV companies were christened: Gridtronics, Home Box Office, and Teleprompter/Magnavox. The three were joined by Theatrevision, which began operations in December 1972, and Optical Systems, which commenced its operation in March 1973.

Legal position

Pay-television development is highly influenced by the many, complex FCC regulations to which it must conform. In summary the regulations with the greatest impact state that pay-TV may not show movies which are between 2 and 10 years old and may not carry serials or advertisements. At least 10 percent of the pay-TV hours must be devoted to programming other than movies or sports. At present the types of films which may be shown are unrestricted, although most cable operators have avoided X-rated films.

Industry dilemma

Two methods of charging for pay-TV service—per channel and per program—are considered feasible. With the per-channel method the subscriber is charged a fixed monthly fee for the pay-TV service. With the per-program format the pay-TV subscriber orders and is charged for each program individually. The per-channel system costs the pay-cable entrepeneur less than a per-program system, primarily because of the difference in equipment costs. A per-program system offers an opportunity for higher total revenues, however, because it is tied directly to the amount of programming watched. Most of the pay-cable companies in operation at this date bill on a per-channel basis.

Industry progress

According to a National Cable Television Survey, as of April 1974 forty-six pay-cable systems were operating in the United States.[3] These systems served more than 500,000 CATV subscribers, 60,000 of whom subscribed to the pay-cable services. At that time, pay-cable was being offered in Arkansas, California, Florida, Georgia, Massachusetts, New Jersey, New York, Ohio, Oregon, Pennsylvania, and Virginia. Pennsylvania had the greatest number of systems, twenty; New York followed with a total of nine.

All of the systems offered six to eight feature films monthly. Half of the pay-cable systems offered films and sporting events on a regular basis, along with some hobby, travel, and cultural programs. The balance of the systems offered feature films only, but planned to introduce sports, educational, cultural, and other programming in the future.

Although only 60,000 subscribed to pay-cable in the 46 areas served by pay-cable in 1974, the impact was apparent. The National Association of Theatre Owners strenuously opposed pay-cable in the courts, and the National Association of Broadcasters established an "anti-pay" committee to raise $500,000 with which to fight the growth of pay-cable. Both associations fear their audiences will diminish if pay-cable becomes popular nationwide.

The National Cable Television Association has formed its own pay-cable committee to promote the industry's growth. It has also allocated considerable time to planning its national convention programs to expose cable operators to the problems and opportunities which pay-cable affords.

Characteristics of Four Pay-Cable Services

In view of the interest in pay-cable television, it is pertinent to explore the operations of four pay-cable entrepreneurs in order to depict the versatility of this service. These four systems will be analyzed from the point of view of the cable operator who, interested in offering to his subscribers premium entertainment, is in the process of deciding which pay-cable service would best meet his requirements. Two of these companies, Optical Systems and Home Box Office, are currently in operation and are expanding. The other two, Theatrevision and X-TRAA-vision, are not. They do, however, offer interesting insights into ideas which, although they have presently failed, may be resurrected in the future.

Optical Systems

Optical Systems began operations in San Diego, California, in March 1972. At one point it was signing 500 subscribers per week from its potential market of 70,000.[4] The subscriber fee included a $20 deposit for the per-program converter, $6.00 for the first six months' maintenance, and $6.50 for the first month's films.

The subscriber receives pay-TV programming by purchasing tickets at specified stores and offices or by mail. The key-punched ticket is inserted in the set-top converter, which reads the punches on the ticket, allows the television circuits to receive the pay-TV program, and destroys the card.

Optical Systems offers a variety of ticket arrangements. For instance, a weekly ticket costs $2.25 and is good for seven days' unlimited use for any movies shown on the pay-cable channel; a monthly movie pass is $6.50, and a 5-game package of televised San Diego Conquistador basketball games is $7.50.

The September 1973 movie offerings included "Stand Up and Be Counted," "Nicholas and Alexandra," "Rage," "Steelyard Blues," "The Train Robbers," "The Thief Who Came to Dinner," "To Find a Man," and "1776."[5]

Optical Systems paid for all expenses associated with acquiring the movies, purchasing the distribution equipment, and marketing their product. The owner of the cable system received 10 percent of the gross revenues.

In Toledo, Ohio, Optical Systems uses a per-channel concept (as they plan to do in any new system they open). They began operations on September 6, 1973. By the end of January 1974 they had 6,345 pay-cable subscribers out of a potential 25,000. An estimated 200 of these pay-cable subscribers had not previously been cable subscribers.

The programming is the same in Toledo as in San Diego with the addition of sporting events distributed by North American Cable. The fee to the subscriber is $6.50 per month, plus a $1.50 maintenance charge and a $25 deposit for the converter. The converter, which uses mid-band frequencies, operates similarly to a conventional television tuner. No tickets are required to activate the channel.

Optical Systems assumes all costs associated with delivering programming to the pay-cable subscriber. The cable operator receives approximately 10 percent of the gross revenue.

A company executive stated that all new Optical Systems operations will be developed in a three-step process: enter with a single channel per-channel format, add a second channel of per-channel programming, and finally add a channel of per-program programming. He also said Optical Systems is in a position to expand its service to other cable systems but the minimum cable system size the firm presently will consider is 20,000 subscribers.[6]

Home Box Office[7]

Home Box Office is a subsidiary of Sterling Communications, which in turn is controlled by Time-Life. Home Box Office is presently operating on a per-channel pay-cable approach in the Pennsylvania communities of Wilkes-Barre, Palmerton, Lansford, Ironton, Allentown, Bethlehem, Hazelton, and Mahanoy City, and in the New York communities of Islip, Mount Vernon, Ithaca, Vestal, and Jericho.

In addition to first-run movies, Home Box Office offers the Knicks and Ranger games from Madison Square Garden, college basketball games, track meets, dog shows, roller derby, etc. Also included in the fare are intermittent afternoon specials for the children and Sharon Obek's introduction of the nightly movies. Movies offered in September 1973, included "Young Winston," "Living Free," "Cancel My Reservation," "Rage," and "The Train Robbers."

Pay-cable subscribers usually pay $6.00 per month for the programming service. However, the cable operator has the right to set the monthly fee. Home Box Office receives $3.50 of the first $6.00 and 50 percent of any charge above $6.00. In contrast to Optical Systems, Home Box Office does not incur all the associated expenses. They do provide the programming and pay for the cost of distributing the programming by common carrier (usually microwave) to that point which passes closest to the cable system's head end. But the cable system is responsible for selling the service, purchasing and servicing the set-top converters, marketing the system, and collecting the bills. Home Box Office will furnish marketing advice and provides monthly program guides which the cable operator can purchase at cost (not to exceed ten cents per program).

Home Box Office has the ability to expand. Judging from their history they provide this service to cable systems of all sizes.

Theatrevision[8]

Until February 1974, Theatrevision was one of the few per-program pay-cable companies in existence. It began operations in December 1972, with a test of 1,000 homes in Sarasota, Florida.

The operation used Monitron set-top converters which were activated by ticket. One pay-channel with three programs (usually movies) to choose from nightly was offered. Special sporting events such as the Foreman-Frazier heavyweight fight were also shown. September 1973 movies included: "Happy Birthday, Wanda Jean," "Day in the Death of Joe Egg," "You'll Like My Mother," "The Thief Who Came to Dinner," "Treasure Island," and "Up the Sandbox." Each movie cost $2.00. Tickets were available at supermarkets, drugstores, vending machines, and the cable television office.

Subscribers paid a $20 deposit for the converter, a $6.00 fee for installation, and $14 for seven movie tickets. Theatrevision assumed all the costs associated with providing the pay-cable service; in return the Sarasota cable operator received approximately 10 percent of the gross revenues.

At the NCTA Convention in June 1973 Theatrevision announced plans to offer their pay-cable service in six additional cable systems with a total of 170,000 cable subscribers, but these plans were never realized.

In February 1974, Theatrevision announced it was suspending operations in Sarasota and was cancelling plans to expand. The reason given was technical problems with their original set-top converter. At this time the cost of a new set-top, per-program converter was considered too high to provide an adequate return on investment. (It is not known at present whether Theatrevision will resume operations in the future.)

X-TRAA-Vision

Although X-TRAA-Vision had not begun operations at the time of our research, the firm's unique ideas warrant consideration in this review.[9] As of 1973 this pay-cable company was to begin operations in Carrolltown, Pennsylvania, using a per-channel format. The subscriber fee was to be $6.00 per month.

The pay-cable system was geared to the needs of the small cable operator. X-TRAA-Vision was to assume all costs associated with providing the subscriber with pay-programming. In return for the use of a channel on the cable system, the cable operator could choose to receive his share of the revenues in one of two ways: (1) The gross monthly revenue could be split 50–50 after deduction of a pro-rated converter cost, programming costs, pro-rated video delivery cost, and local administration costs; or (2) the cable operator could receive a monthly fee per subscriber—$.30 for the first 250 subscribers, $.45 for up to 500 subscribers, $.60 for up to 750 subscribers, $.75 for up to 1,000 subscribers, and $.90 for every subscriber over 1,000. Subscriber contracts were to be for three years. A one-year trial period was to be available if the cable subscriber wanted to evaluate the system. During this trial year 20 percent of the cable system's monthly pay-cable receipts were to be held in escrow. If the contract were terminated at the end of the first year, the funds would revert to X-TRAA-Vision to defray the costs of the converter. Otherwise the escrow was to be released to the cable operator. At the end of the second year, 10 percent of the receipts were to be held in escrow. During the third year there would be no withholding.

The Carrolltown, Pennsylvania, cable system was to be served by X-TRAA-Vision, beginning July 1973. A phone call to the cable system operator in March 1974, however, revealed that although the plans had been made to begin operations, he had not heard from the company for the last eight months. Their phones had been disconnected and he had no idea where to get in touch with the company.

Concluding Comments

The cable operator has a strong incentive to offer pay-cable service to his subscribers. Pay-cable may increase his average revenues per subscriber, and it may provide a competitive edge in inducing prospects to sign up for cable service as well.

Since a number of firms are available to provide pay-cable services for the system operator, adding this service is readily facilitated. But the cable operator must decide which pay-cable supplier to contract with and how much of the work involved in offering premium entertainment he wishes to do himself.

Of prime importance, of course, is the kind and quality of entertainment. While television viewers may favor pay entertainment made available without commercial interruptions, they will not be attracted by mediocre programs. The cable operator's task, therefore, is to evaluate the kind and quality of entertainment available from competing pay-cable suppliers.

The operator must also consider the quality, as well as cost, of the converter furnished by the pay-cable supplier. Unless operation is trouble-free, subscribers may become disenchanted and cancel the service. Information on the performance of hardware in operating situations is of key importance to the cable operator in evaluating alternative pay-cable suppliers.

In addition he must weigh:

1. Conditions influencing the cost of pay-cable service to the subscriber—such as the amount of the deposit for the converter, maintenance costs, and the monthly entertainment

charge. Unless the prospective subscriber perceives these expenses as attractive compared to hiring a babysitter and going out to view the entertainment, he will not sign up for pay-cable service.

2. The convenience with which the premium entertainment may be purchased. (On this point it would appear that the per-channel charge may have an advantage over the per-program method.)

3. The marketing assistance provided by the pay-cable supplier. When pay-cable is introduced effective marketing effort is required to inform prospects of the new service and to encourage them to purchase it. Help in follow-up marketing efforts may also be important to the cable operator in getting enough users to make pay-cable a profitable venture. Hence, careful evaluation of the quality of and charges for the marketing services provided by alternative pay-cable suppliers is another key step in the decision-making process.

Analysis should also include a comparison of the revenue-sharing arrangement with the alternative pay-cable suppliers under consideration as well as of the services and associated costs that they will assume under the contract.

According to the National Cable Television Association survey, there are three basic financial options available to the cable system operator:

1. A joint venture in which the pay-TV entrepreneur provides programming to the cable system operator, who assumes responsibility for capital equipment costs, marketing, billing, and technical functions. In 1974 twenty-two systems reported that they were using this approach.

2. A lease arrangement in which the pay-TV entrepreneur leases a channel from the cable operator and assumes, in addition to program procurement, all responsibility and costs for capital, equipment, origination, converter installation, billing, and marketing. As of 1974 eight systems were using this approach.

3. An independent venture in which the entrepreneur owns and operates both the cable systems and the pay-TV service. Under this arrangement, the pay-TV entrepreneur/cable system owner provides the pay programming and hardware to his cable system, which in turn handles converter installation, hardware maintenance, marketing, and billing functions. In 1974 thirteen systems were using this approach.[10]

NOTES

1. Adapted from *Videocassette & CATV Newsletter* 4, No. 7 (April 1974):9-10.
2. *Pay TV Newsletter,* Feb. 8, 1973 [Paul Kagan Associates, Inc.].
3. Adapted from *Videocassette & CATV Newsletter* 4, No. 9 (June 1974):13.
4. Except where noted otherwise information in this section is based upon interviews with executives of Mission Cable TV, San Diego, in June 1973, and the Resident Manager of the Optical Systems operation available over Buckeye Cablevision, Toledo, Ohio, in February 1974.
5. *Pay TV Newsletter,* September 10, 1973.
6. Interview with Allan Greenstadt, Marketing Dept., Optical Systems, Inc., Los Angeles, February 1974.
7. Information on Home Box Office operations gained through personal interviews and correspondence with executives of Home Box Office in November 1973.
8. This section based upon presentation made by a Theatrevision executive at NCTA Convention, June 1973, supplemented by interviews with the general manager of the firm during January and February, 1974.
9. The section on XTRAA-Vision is based on a news article in *Video Publisher,* June 13, 1973, information contained in company literature, and an interview with the manager of the Carrolltown, Pa., CATV System, Mar. 1974.
10. *Videocassette & CATV Newsletter* 4, No. 9 (June 1974):13.

IV

LOCAL PROGRAM ORIGINATION

Policy of Localness

In 1969 the Federal Communication Commissions (FCC) issued a rule requiring cable television systems with 3500 or more subscribers to originate local programming "to a significant extent." This rule (FCC Docket 18397), later upheld by the U.S. Supreme Court, follows the long-standing FCC policy of promoting "localness" in programming.

The basic rationale for local programming stems from the phrase "public interest, convenience, and necessity," embodied in the Federal Radio Act of 1927. In the Federal Communications Act of 1934 the phrase was repeated along with a more specific reference to localness in Section 307(b), which required the FCC "to provide a fair, efficient, and equitable distribution of radio service...." Section 307(b) has been interpreted as requiring not only an equitable allocation of broadcast frequencies but also a local program service and community access to broadcast facilities. The Supreme Court affirmation of this interpretation said, in part, "Fairness to communities is furthered by a recognition of local needs for a community radio mouthpiece."[1]

The concept of local needs was further defined in the FCC's 1960 *Report and Statement of Policy on Programming.*[2] The 1960 statement listed fourteen "major elements usually necessary to meet the public interest, needs, and desires of the community." The first element, "the opportunity for local expression," and most of the other categories depend, at least in part, on the local circumstances. Thus the FCC's chief criterion for broadcast program evaluation is the concept of "localness" defined as serving the needs of the local community.

NOTE: This chapter was prepared by Darrell Dahlman, Research Associate, Graduate School of Business Administration, The University of Michigan, under the direction of the author.

In 1966 the FCC went a step further by requiring broadcasters to ascertain the program needs of their communities and, as the method of ascertainment, to list on station renewal applications the public needs and interests and the programs planned to satisfy those needs. Clearly the policy of the FCC is that "a station serves its area as a means of community self-expression, giving it a broadcast voice as well as a broadcast ear."[3]

The FCC policy of localness for broadcast television was implemented in the 1952 FCC rules. It soon became clear, however, that allocation of frequencies could not be equitable and the concept of genuine localness was impossible within the limitations of the frequency spectrum. As Ralph Lee Smith points out:

> In addition to discrimination against the large percentage of the population beyond the clear signal range of a few major metropolitan centers it [broadcast television] has not been able to offer real community service or community expression. When off-the-air broadcasters say that they provide a community or local signal, they mean that they send out a highly generalized service to a large, poorly defined audience.[4]

Local Origination and CATV

True localness cannot be obtained by broadcast television, but cable television by its very nature is an ideal means of community expression. With the rapid growth of CATV during the 1960s, it was evident that cable could provide a unique supplement to television by originating local programming. The FCC recognized CATV's potential in this area with its 1969 local origination requirement and again in the 1972 CATV rules, which also stated the basic objective as "getting cable moving so the public may receive its benefits...."[5]

Section 76.5(aa) of the CATV rules defines origination cablecasting as "programming (exclusive of broadcast signals) carried on a cable television system over one or more channels and subject to the exclusive control of the cable operator." By the FCC's broad definition, origination programming need not be truly local in origin. Programs may be produced within the cable system's studios or at least within the community, but syndicated programs or other programs from outside sources can also be considered as locally originated.

It is obvious from the FCC broadcast television rules that the main program criterion is not so much where the programs are produced as whether the programs meet local needs. No doubt many community needs or interests are unique to situations or circumstances within the CATV system. But many other needs also are common to numerous CATV systems, as indicated by a collection of ideas and aids for local origination published by the

National Cable Television Association (NCTA).[6] This guidebook
covers eleven program areas, giving hundreds of program ideas
suitable for different communities. Many of these programs need
not be produced within a cable system community. Special programming could be used in areas having a substantial audience
with similar interests. Such audiences include ethnic or religious
groups, professionals, senior citizens, hobbyists, and many others.
Significant programming for special interest audiences too small to
be served by broadcast television is possible on cable, and the
programs need not be produced locally.

By offering local origination to their subscribers, CATV systems
can serve purposes in addition to that of fulfilling the FCC
requirement. This point is demonstrated by the fact that 40 percent of the systems presently originating have fewer than 3,500
subscribers[7] and thus are not subject to the FCC rule.

One reason for local origination is building subscribers. Programming which contains information of local interest, especially
if it is unobtainable on broadcast television, can serve as an
incentive for CATV subscription. Furthermore, locally originated
programs, if well done, can draw audiences large enough to make
advertising feasible. This could become an important source of
revenue for cable systems. In addition, the good public relations
benefit of local origination reinforces the image of CATV as truly
serving local needs and interests.

The problem

As the 1973 NCTA *Local Origination Directory* states, "There
is little, if any, profit in local origination; advertising never quite
foots the bill."[8] Projected costs for full local origination vary
considerably. The FCC estimates that on an average CATV system
(4,500 subscribers) local origination production in color would
require about 35 percent of the system's subscriber revenue.[9]

Obtaining much origination programming from outside sources
is one means of lowering costs. These programs are available from
various free sources, private or governmental, and from commercial program distributers. Other sources include individual CATV
systems and state or regional cable associations.

Regardless of the source, it is probable that some CATV systems
will not be able to afford a complete origination schedule. As the
1973 NCTA Directory indicates, only 514 of the cable systems in
the United States have local origination,[10] and those systems with
origination only average 21.1 hours of programming a week.[11]
That figure may be misleading because of the practice of repeating
programs throughout a week.

Because of the financial and organizational problems inherent in local origination, the most substantial progress in this activity must be expected from Multiple System Operators (MSOs). MSOs should have the economic base required to produce or obtain quality programs, and potentially they can use programs on a substantial number of systems within their control. An MSO organization provides a ready-made system of distributing programs.

Survey of MSOs

A number of questions concerning MSO programming arise. How many hours of program material do MSOs originate each week? How much origination is done by individual CATV systems for use by other systems within an MSO? What independent sources of programs are used by MSOs? How are programs distributed within an MSO? What determines the selection of programs? What interconnection capabilities are being used? What do MSOs plan for future programming?

A basic hypothesis of this study is that MSOs should be expected to produce significantly more programs per week than independent CATV systems. They should utilize many different sources for programs, including their own CATV systems, and there should be little, if any, interconnection of systems at this time although most MSOs should plan interconnection in the future.

Method

To determine the answers to these questions, a number of large Multiple System Operators were sent questionnaires during fall 1973 (see Appendix A). The nine responses received represent MSOs controlling from five to 140 individual CATV systems. Total subscribers for each MSO ranged from 20,000 to 800,000. Thus, the nine MSOs should provide a cross-section of the industry. The survey involved 426 individual CATV systems with over two million subscribers.

Results

The majority of the respondents—seven of nine—produced programs for their systems; the amount produced varied considerably. Average weekly production ranged from twenty to 364 hours, with a mean of 84.8. (The mean is distorted, however, by the largest amount. Eliminating that figure gives a more representative mean of 38.3 hours per week.)

Five of the nine MSOs indicated that individual CATV systems produced programs for use by other systems in the company. Of these, the production is done by only 55 of the 271 systems. (See Table 8.) Five MSOs obtained programs from independent sources, including a wide range of government, private, and commercial sources. (See Figure 5.)

Table 8
MSO SURVEY RESULTS

Code #	Systems	Subscribers	MSO Origination Hours/Week	Systems Originating	Individual Systems Origination Hours/Week
1	140	800,000	20	42	casual basis
2	88	374,000	364	4	5½
3	32	265,000	...	...	...
4	52	200,000	10 - 45		
5	28	120,472	100	1	N.A.
6	41	85,000	N.A.	...	...
7	30	85,000	...	...	...
8	5	60,000	16 - 24	3	N.A.
9	10	20,000	60	5	3
TOTALS	426	2,009,472	594	55	

The most frequently used method of distribution was bicycling.[12] More MSOs made multiple copies of programs on video tape than on film for distribution to CATV systems. In most cases, programs were selected by the individual system, not by the MSO. Of the nine MSOs surveyed four had interconnection capabilities and six in all intend to have interconnected systems. Projected programming plans include networking by satellite (seven of nine) and cooperation with other MSOs (five of nine).

Discussion

The results of the survey first must be viewed in light of the present development of the CATV industry. The large number of mergers between CATV companies indicates a state of flux at this time. Present emphasis appears to be more on building large, efficient, and profitable organizations than on developing extensive programming capacity. When the industry is more stable more emphasis on local origination can be expected.

Even so, the survey's findings at this time are encouraging. Most individual CATV systems do not originate programs and, of those

Independent Program Sources
Cable Network Television
CAT Video
Telemation
Olympus
Videomation
NTA Films
Tandem Four
Spanish International
Official Industries, Incorporated
Greatest Fights of the Century
Modern TV
Sterling Films
University of Illinois
Christophers
H. E. W.
Social Security

Fig. 5. Independent program sources actually used by respondents to survey of MSOs.

that do, the average is only 21.1 hours per week. Only two of the nine MSOs surveyed do not originate programs, and one of these indicated possible origination in the future. The other MSO would prefer to see its local systems design their own programs for each community. Those MSOs which originate programs do so at an average of 38.3 hours per week. Only 60 of the 514 individual CATV systems originating programs exceed the amount set by the MSOs.

The most surprising results of the survey were those indicating the lack of MSO activity in distributing programs from one system to other systems within the company and the MSOs' failure to utilize central buying and distributing techniques for syndicated programs. The company organization and the simplicity of bicycling programs would make both practices relatively easy.

Although most of their activity is on a limited basis at this time, facilities are being interconnected by four of the nine systems and it is significant that most of the MSOs intend to have interconnection by satellite at a future date.

Recommendations

Origination by MSOs will probably increase in the near future and will also probably stimulate increased origination by independent systems. The most effective means of stimulating local origination, however, would be for the FCC to clarify its definition of local origination and to require some truly local origination. This move, coupled with a requirement to ascertain local needs, would make local origination a service to the community.

Other Local CATV Programming:
Types and Sources of Material Used

Cable television systems present locally originated programming primarily for two reasons. First, there is pressure from the Federal Communication Commission's requirement for origination "to a significant extent" by systems with 3,500 or more subscribers. Second, local origination programming can be used to encourage subscription to CATV.

According to the 1973 National Cable Television Association (NCTA) *Local Origination Directory,* 585 of the nation's 3,000 cable systems now originate programming compared to only 284 in 1971. These 585 systems cover 1,363 communities and reach nearly four million subscribers. Since 40 percent of the originating systems have fewer than 3,500 subscribers, the number of systems originating programs should continue to increase.

The NCTA directory also revealed that the average amount of origination per system was 21 hours per week and that the programming was divided among six program categories.

To find out more about CATV programming, in spring 1973 questionnaires were sent to over 30 companies listed in the "Directory of Television Program Sources and Services," *1973–74 TV Fact Book.* Responses were received from 23 companies, although only nine of them completed and returned the questionnaire. Of the others, eleven indicated either they did not provide programming to CATV or that they are no longer providing such programming. Thus, this report is based on the nine questionnaires returned, plus the brochures or promotional material from six other companies.

The purpose of the study was to determine the types of programming available to CATV and the extent to which various companies offer programming. Obviously, the limited sample does not allow for a definitive examination of the sources of CATV programming. But it does provide the basis of several generalizations or conclusions.

1. For the time being at least, large companies involved in television program distribution (such as MCA and Warner Brothers) appear to have little interest in CATV. The apparent reason is the relatively small market of CATV compared to broadcast television.

2. Most companies active in CATV programming are small, but several of them have extensive saturation of the market (up to 350 systems).
3. The average amount of programming per week provided by each system could be significant. Several companies provide almost one-half of the average amount of weekly programming.
4. The most common format for programs was 16 mm film and ¾-inch video cassette.
5. As might be expected, program types varied considerably. The range of programs included free "sponsored" films, re-runs of old TV series, feature films, and original programs produced specifically for CATV.
6. Also, as might be expected, the cost of programming appears to be in direct relation to the quality of programming. In general the re-runs and old features cost a flat rate of from $20 to $90 per program, while the more exclusive programs were priced on a sliding scale depending on the number of subscribers per system.

Table 9
RESULTS OF CATV PROGRAMMING QUESTIONNAIRE

Company	Form	Number of Systems	Hours/ Week	Type of Program	Cost
Association Sterling	Film	350	10	Documentaries	Free
Cablevision Bingo	Service	200-300	—	Bingo	Up to $1,950
CNT		130	—	Feature films, syndication, cartoons, etc.	$40-$90 per feature
Hemisphere Pictures	16 mm	6	8	*Cisco Kid, Carnival of Terror*	Varies
Official Films	16 mm	60	5	Peter Gunn, etc., plus 50 feature films	Depends on market
Rowland Productions	1″ tape and ½″ cassette	75	4	How-to shows, children's, musical, etc.	$8.50 to $20 per show
Tape-Athon Corp.	Audio tape	144	—	Background music	$125/mo.
Valentino, Inc.	Audio tape	30	—	Production music	$640/pkg.
World Cable Corp.	¾″ U-matic cassette	26	15	Spanish programming	Depends on percentage of Spanish population

Table 10

OTHER SOURCES OF CATV PROGRAMMING

Company	Form	Type of Program	Cost
Modern Talking Picture Service	¾" cassette	RR programs	Free
Modern Video Programming	16 mm film and ¾" cassette	Walt's Workshop, Art of Football, etc.	Up to $225/ series
Red Eye Net	¾" cassette	Five-hour block of entertainment and information programs	Depends on number of subscribers
Tele-Media, Inc.	¾" cassette	Adventure, documentary, and education	$25–$50/program
Time-Life Films	Tape and film	Drama series, entertainment, and public affairs specials	Varies
Video-Mation	16 mm film	Music, sports, interviews	Varies

Note: Compiled from brochures or promotional materials from six companies who did not answer the questionnaire.

NOTES

1. FCC v. Allentown Broadcasting Corporation, 349 U.S. 358 at 363 (1955).
2. FCC, Report and Statement of Policy re: Commission en banc Programming Inquiry, 25 Fed. Reg. 7291 (1960).
3. Sydney Head, *Broadcasting in America* (Boston: Houghton Mifflin, 1965), p. 375.
4. Ralph Lee Smith, *The Wired Nation* (New York: Harper & Row, 1972), p. 14.
5. FCC, 1972 Rules Par. 58.
6. NCTA *Cablecasting Guidebook* (Washington, D.C.).
7. NCTA *Local Origination Directory* (Washington, D.C., 1973), p. 85.
8. *Ibid.*, p. 1.
9. Martin Seiden, *Cable Television U.S.A.* (New York: Praeger, 1972), p. 31.
10. NCTA *Directory*, p. 11.
11. *Ibid.*, p. 85.
12. In bicycling, program material scheduled for use by several CATV stations on specified dates is sent by the originator of the program to the first station on the list and thereafter is delivered from one using station to the next without being returned to the producer of the program after each showing.

IV: APPENDIX A

MSO PROGRAMMING SURVEY

1. Total number of cable systems over which you have operating control. _____
2. Total number of subscribers in the above systems. _____
3. Program origination:
   A. Does your company originate programs for the CATV systems you control?   (Please circle)   YES   NO
   How many hours per week? _____
   B. Do individual CATV systems controlled by your company originate programs for other systems in your organization?   YES   NO
   How many CATV systems do this?
   How many hours per week? _____
4. Do you obtain programs from independent producers or distributers (e.g. syndicators) for use throughout your system? YES   NO
   A. If "yes," please list the names of these sources.

   _____
   _____
   _____
   _____

(To be detached) -

NAME_____ CODE NO._____

TITLE _____

COMPANY_____

ADDRESS _____

                                              (Please check
5. Methods of program distribution: appropriate items)
   A. Physical arrangements:
      1. Bicycling   _____
      2. Multiple Copies:
         a. Video tape   _____
         b. Film   _____
         c. Other   _____

B. Selection of systems:
    1. Each system chooses which programs to use   _____
    2. Company requires use by some systems   _____
    3. All systems use the programs   _____
    Comments:

_____
_____
_____

6. Do you have interconnection capabilities among any of your systems?   (Please circle answer)   YES   NO
Do you plan to interconnect any systems?   YES   NO
Comments:

_____
_____
_____

7. What is the nature of your projected plans for programming? (Please check appropriate items.)
A. Cooperation with other MSOs   _____
B. Utilization of a satellite for networking   _____
C. Arrangements with independent TV stations   _____
D. Other (please specify)   _____

_____
_____
_____
_____

IV: APPENDIX B

CATV PROGRAMMING QUESTIONNAIRE

1. Program form offered to CATV systems: (Please Complete)
   Type of Film _____
   Type of Tape _____
   Other _____
2. Number of CATV systems using programs or service:_____
3. Average hours/week per system occupied by your program
   material: _____
4. Please provide a list of all programs or services offered to CATV
   systems.

5. Please furnish a price list for the programs or services offered.

6. Name and address of your firm; name and title of official complet-
   ing questionnaire.
   _____
   _____
   _____
   _____

V

TOWARD THE DEVELOPMENT OF CATV PROGRAM NETWORKS

Related to the problem of making attractive program material available on the cable system's own channels is the possibility of developing program networks through which high quality material developed by one CATV operator may be made available to CATV systems in other areas. While wider use of such programs would enable the producer to invest more money in the development of his material and use the° promise of production of high quality programs to attract sizable viewing audiences to cable system channels, the possibility of interconnecting cable systems via microwave or satellites adds an even greater dimension to CATV potential.

An additional development which has much promise is the organization of regional cable systems networks centered around outstanding independent over-the-air broadcasting stations. The possibilities of these developments are considered in this chapter.

Interconnection of CATV Systems via Satellites[1]

Introduction

The desire of man to communicate over great distances is responsible for the technology which made the telegraph, telephone, radio, and television possible. The nature of its signal made radio a natural means of instantaneous communication between almost any two points on earth. Because its wide band

NOTE: This chapter was prepared by Darrell Dahlman, Research Associate, Graduate School of Business Administration, The University of Michigan, under the direction of the author.

signal requires the use of higher frequencies, television, on the other hand, was limited essentially to local communication (unless its transmitters were interconnected by microwave or coaxial cable)—until the advent of satellites made global television also a reality.

Communications satellites provide an improved means of sending signals (telephone, telegraph, radio, or television) over vast areas. Their development has greatly changed the process of intercontinental and intracontinental communication during their relatively brief history, and it is clear that communications satellites will produce even greater changes in the near future.

History

The United States began the era of space-age communication in 1958 with the first successful experimental communications satellite—Project SCORE. The early experiments which proved that satellite communication was practical, including Echo in 1960, had severe disadvantages. The main disadvantage was design—all of the early satellites were passive, which meant they could only reflect signals back to earth. This design also meant an inefficient use of transmission power and only permitted the relay of telephone, facsimile, and data signals.

These problems were quickly overcome with the Courier, Telstar, and Relay satellites of 1961–62, which carried equipment for the reception, amplification, and retransmission of earth signals. The "active" satellites proved the feasibility of using satellites for intercontinental communication including television. They did have one major disadvantage, however. Their orbits were nonsynchronous—i.e., they moved in relation to the earth's orbit—which meant that as these satellites circled the earth the amount of time usable for communications was often limited to only a few minutes.

Syncom I solved that problem by using an orbit of 22,300 miles, thereby remaining synchronous, i.e., remaining in the same location relative to the earth. Transmission from a synchronous satellite can cover approximately one-third of the surface of the earth. Thus, only three satellites are needed to cover the entire earth.

By 1962, therefore, the technology for global communication was in existence. Next an organization to administer it and to provide commercial satellite service was needed. COMSAT, the Communications Satellite Corporation, was created by Congress in 1963 to fill that need. COMSAT also represents the United States in INTELSAT, the International Communications Satellite Consortium which owns the satellite system.

With the launching of Intelsat I in 1965, satellite communication reached full operational status. Since then, three additional Intelsat series have been launched. The latest, Intelsat IV, provides efficient intercontinental communication service and is far superior to the earlier Intelsat series, as Table 11 illustrates.

Domestic satellites

The success of Intelsat with international communications created much controversy in the United States regarding satellites designed specifically for domestic use. In 1965 the American Broadcasting Company proposed a network-owned system to be used by the networks for radio and television services. Since the Communications Satellite Act did not prohibit other U.S. satellite systems and there was no established FCC policy, in 1966 the FCC asked for additional proposals on the subject.

In 1969, the President's Task Force on Communications Policy recommended that a pilot domestic satellite system be authorized to determine the cost and demand for such services. The recommendation was ignored.

Meanwhile, Canada became the first country with an operating domestic satellite system. In 1972 NASA launched Anik I for Canada. Anik I is a twelve-repeater system with channels available for rent at $2–3 million per year. The system can handle ten color TV channels or 9,600 telephone signals simultaneously.

Finally, in 1972, after six years of indecision, the FCC adopted an open-entry policy for domestic satellites. Essentially, the decision meant that virtually any company with enough capital could enter the domestic satellite business. In 1973, the FCC approved the plans of six companies to enter the field. The six included: Western Union, Comsat, American Telephone & Telegraph, American Satellite Corporation, RCA Global Communications, and GTE Satellite Corporation with National Satellite Services, Incorporated.

The obvious effect of six competitors' vying for the domestic communication market is a rush to find customers, yet the size of the potential market is unknown. One estimate for data transmission alone is that the present market of $1 billion a year will increase to nearly $5 billion by 1980. In addition, there is the telephone and television potential.

In 1973 some experts predicted that not all of the companies would attract enough customers to succeed. One company official flatly predicted that no more than three of the six would survive

Table 11

DEVELOPMENT OF INTELSAT
COMMUNICATION CAPACITY

Description	Intelsat (Satellite)			
	I	II	III	IV
Year	1965	1966–67	1968–69	1971–73
Capacity:				
voice or TV	240 or 1	240 or 1	1,200 or 4	6,000 or 12
Cost ($ million)	4	4.5	6	13.5
Circuits	240	240	1,200	6,000
Design lifetime (in years)	1.5	3	5	7
Circuit years of capacity	360	720	6,000	42,000
Investment per circuit year of capacity (in dollars)	15,300	8,400	1,450	500

the first few years. One suggested alternative was that companies rent space on existing satellites instead of launching their own.[2]

By February 1974 the first of the six companies dropped out of the running. American Satellite Corporation announced it would abandon plans to send up a satellite and would instead lease space from Western Union beginning July 1974.[3] Western Union Telegraph Company, the first company to be approved for a domestic system, was the first in operation with its own satellites. The company launched two Westar satellites in 1974, beginning commercial operations in July. It is estimated that the Western Union system cost about $70 million.

Of the other companies involved in domestic satellites, RCA was the first to begin service, although not on its own satellite. RCA leased space on Canada's Anik II, launched in April 1973. By building four ground stations RCA managed to begin service in January 1974. As of April 1974, RCA services included transmission of all forms of electronic information—voice or image, message or data—from the East and West coasts and to Alaska. Charges for the service are reported to be considerably lower than charges for use of land lines; a three-hour transmission cost $3,500 compared to $6,000 via land lines. Since RCA does not expect to have its own satellites in operation until early in 1976, the firm also leased four channels on the Westar satellites when this service became available.

Uncertainty about demand, the tight money market, and the attitude of the Federal Trade Commission to certain proposed joint ventures, has modified the plans of other competitors. It appears likely, however, that other satellites will eventually be placed in operation to serve both commercial and public interests.[4]

Implications for CATV

The use of satellite transmission for commercial television will probably not create much change, with the exception of its lowering costs for interconnection. The real beneficiary of satellite transmission would be cable television. Using a channel on a domestic communications satellite, CATV systems could interconnect to form a network of all 3,000 cable systems. With the audience which would result, CATV could provide the economic base for advertising revenue or pay-TV. This development could in turn produce substantial programming alternatives.

In 1974 the Cable Satellite Access Entity, a consortium of some 40 CATV companies, financed a feasibility study by Booz, Allen, and Hamilton to investigate this possibility. The study was designed to determine CATV market areas which would be served, total communications capability of the system, number of simultaneous channels required, number of hours of service required daily, and the overall cost of such a system. As of this writing, there has been no public announcement of the findings of the Booz, Allen, and Hamilton study.[5] The technology for a cable satellite system exists today, however, and the only barriers to the implementation of the system are the financial arrangements and the policy agreement.

As this report nears completion it is announced that Home Box Office will supply approximately 80 percent of Teleprompter's cable system subscribers with pay programming via domestic satellite in the fall of 1975. This move will capitalize on the potential benefits to come from national distribution via domestic satellite of attractive pay cable programming. Teleprompter officials expect that the provision of pay cable program service to its subscribers should make a significant contribution to revenue and net income.[6]

Analysis

Domestic communication satellites could revolutionize cable television. In its current fragmented state, CATV does not have the economic power to compete against other media. With satellite interconnection, cable television can become a viable alternative, capable of producing a wider range of programming without the need to program for mass audiences. A cable satellite system could provide programs from a diversity of sources to a diversity of audiences, thus fulfilling the Sloan Commission's call for "television of abundance". The plans by Home Box Office to distribute pay cable programs via satellite are a significant example of the possibilities.

Development of Regional CATV Program Networks
Centered around Independent TV Stations

One of the main reasons for subscribing to cable television is the availability of more broadcast television stations, particularly since the FCC cable rules allow CATV to carry independent TV stations in addition to the TV network outlets. While CATV systems can use their carrying of independents to attract new subscribers, the independent stations can also use cable to build their audiences. Cable television systems and independent television stations can, therefore, have a symbiotic relationship.

In order for the relationship to be beneficial, both sides must exploit the potential advantages. Cable must "sell" the additional programming to prospective subscribers, and just as importantly it must constantly remind present subscribers of the variety of programming available only over CATV. Viewers are more likely to maintain their CATV subscription if they know they are getting something otherwise unobtainable. Independent stations, by the same token, must promote the availability of their signals on CATV systems. A TV station has a limited signal range, but cable can extend that range even to a number of different states; cable can make a local station into a regional station.

In most cases an independent is at a disadvantage in competing against network affiliates. Affiliates draw larger audiences, can charge higher advertising rates, and can, therefore, afford better programming. But if the independent TV station is carried by enough CATV systems, that disadvantage can be limited. To use

CATV to an advantage the independent station operator must know how many and what types of people watch his programming on cable, and then use that information to sell advertising.

To determine how effectively independent stations use CATV, in the fall 1973 survey questionnaires were sent to thirty major independent stations in various parts of the United States. Returns were received from fourteen stations, or nearly 50 percent of those sampled. The initial sample consisted of twenty-four members of the Independent Television Association (INTV) plus six other large stations. (There are approximately ninety independent stations in the country, forty-four of which belong to INTV.) The survey results, while not necessarily representative of the entire industry, indicate conditions among the largest independent stations as of fall 1973. (See Appendix A for a tabulation of responses.)

The results indicated a wide range in the number of CATV systems carrying the signal of each independent station and also in the total number of CATV subscribers potentially available to the independent station. The number of systems ranged from 50 to 204, with 109.5 as the mean. Total subscribers ranged from 114,000 to 886,355 with a mean of 339,739.

Obviously, by themselves, these figures do not give the full picture of the independent's cable viewers. More research information is needed to make the data meaningful. For example, some of the CATV subscribers also may be able to receive the station over the air. Thus, the subscribers do not necessarily represent an increase in the total audience of the independent station. Data are needed to determine where the subscribers are located, but, more significantly, information is needed about which subscribers view which programs.

To obtain such data requires audience research. But the survey showed that ten out of fourteen independent stations were not conducting any. Two remaining stations did not respond to the question. Thus, only two of the stations surveyed indicated any specific research of their CATV audience. Other stations indicated that they use standard Nielsen and ARB surveys, but these surveys do not produce separate data on CATV subscribers. Interestingly, though, one of the two stations that researched its CATV audience reported the results showed viewing of the independent station was higher via CATV than over the air.

This finding underscores the need for research. With independent stations reaching from 114,000 to 886,355 homes via cable, a significant number of viewers may be tuning in without the station's knowledge. Lack of research may, therefore, have an important economic effect on the stations.

Since television advertising rates are based on the size or potential size of audiences, some stations may be underpricing themselves. The survey results showed that none of the stations changed their advertising rates because of their additional CATV audience.

As one respondent indicated, meaningful audience research is very expensive, which is probably the reason for its almost total absence. Nevertheless, without it stations may deprive themselves of revenue.

The lack of research also makes it difficult to judge the total impact of CATV carriage. Only one station estimated the increase in audience due to cable. It estimated a 20 to 25 percent increase, but this was without audience research and thus the figure is questionable.

The lack of research may also indicate station managers' lack of concern about CATV carriage. Only three stations bothered to actively promote their broadcast over CATV. Their methods of promotion commonly included letters to new systems, co-op ads in local newspapers, and personal contact. Some stations also distribute program schedules, but the wide majority of stations did nothing to promote their CATV availability.

It is extremely doubtful, therefore, that independent television stations are deriving all the potential benefit from being on cable television. If independent television stations are to benefit from future cable developments, such as satellite interconnections, it is evident that they first must learn more about their own coverage and secondly must cooperate more with cable operators.

NOTES

1. This section is based upon background gained by the author in a course in International Broadcasting, correspondence with six firms involved in building and operating satellite systems, and articles from trade papers referred to below.

2. "Now, 'Domsat,'" *Newsweek*, Jan. 21, 1974, pp. 83–84, and "Five Domsat Firms Get FCC Approval to Head for Space," *Broadcasting*, Sept. 17, 1973, p. 48.

3. "ASC Decides to Fly Later," *Broadcasting*, February 25, 1974, p. 58.

4. *Video Publisher*, May 14, 1974, p. 3, and *Wall Street Journal*, September 30, 1974, p. 5.

5. Adapted from "Cable-Satellite Group is off the Launch Pad," *Broadcasting*, Sept. 3, 1973, p. 27. "Study to Evaluate Use of Satellites for Cable," *Broadcasting*, Jan. 14, 1974, pp. 30–31. These reports supplemented by a letter from a spokesman of Comsat General, Feb. 25, 1974.

6. "HBO Pay-Cable Service to Reach 800,000 TPT Subscribers," *CATV*, June 9, 1975, pp. 5–6. Published by Communications Publishing Co., Englewood, CO 80110.

V: APPENDIX A

RESULTS OF SURVEY OF INDEPENDENT TV STATIONS

Station Number	CATV Systems	Total Subscribers	Promotion	Research	Audience Increase (Percentage)
1	204	600,000	No	None	...
2	97	516,000	Yes	ARB	...
3	131	321,015	No	None	...
4	69	212,221	No	None	...
5	200	180,000	Yes	VPI	...
6	77	159,394	No	None	20-25
7	50	150,000	Yes	None	...
8	74	114,000	No	None	...
9	120	N.A.	Yes	None	...
10	92	408,000	Yes		
11	150	886,355			
12	65	491,800	No	None	...
13	150	200,000	No	None	
14	54	177,828	No	None	...
TOTAL	1,533	4,416,613	5 Yes	2 Yes	
	$N = 14$	$N = 13$			
Mean	109.5	339,739			

V: APPENDIX B

SURVEY OF INDEPENDENT TELEVISION STATIONS

Call Number⎯⎯⎯⎯⎯⎯⎯
Name⎯⎯⎯⎯⎯⎯⎯
Address⎯⎯⎯⎯⎯⎯⎯

1. Number of CATV systems carrying your signal. ⎯⎯⎯⎯⎯⎯⎯

2. Number of subscribers on those systems.⎯⎯⎯⎯⎯⎯⎯

3. Do you address promotion to cable operators to get them to carry your signal?   YES   NO (Please circle)
   If YES, what type of promotion is used?

   Please enclose a copy of your promotional material if possible.

4. What audience research has been conducted on the CATV viewers of your station?

   What organization conducted the research?

   Please enclose a copy of a recent audience study if possible.

5. Based on such research, how much does carriage of your signal on CATV increase your audience?

6. What effect does CATV carriage have on your advertising rates?

   Please enclose a copy of your current advertising rate card.

VI

VIDEO CASSETTE: A COMPETITIVE THREAT?

Since some of the services which CATV systems can render may also be offered by competing alternative media, the video cassette should be examined as a development of substantial potential importance to the CATV market. The use of the video cassette for entertainment, educational, and business purposes will be reviewed briefly with the objective of assessing the extent to which it may become a serious competitive threat to the long-run development of CATV.

The video-cassette industry has been on the verge of rapid expansion for some time. Although observers of the industry predicted a vast market for video recorders or players in homes and businesses with the development of video tape in the 1950s, only in the past several years has the market actually begun to develop. Numerous factors have hampered the widespread use of cassettes, among them lack of standardization and compatibility, high cost, programming shortages, and copyright laws. These technological, economic, and marketing limitations should be overcome in the next decade, however, which will finally give the video cassette wide acceptance. Widespread use of video cassettes is predicted to transform many entertainment, education, and industry habits. It would affect television and CATV programming, motion picture theatres, school curricula, and many other facets of media in our lives.

Technology[1]

Video cassette television is a generic term used to encompass a number of vastly differing systems which allow persons to play

NOTE: This chapter was prepared in April 1974, by Darrell Dahlman, Research Associate, under the direction of the author.

programs of their choice on ordinary television sets. These systems use tapes, discs, film, and other formats, which is why many different technical terms or names exist for the same generic concept.

The proliferation of competing systems, common to emerging technology, finds no exception in the electronics field. Original plans for video-cassette television rested solely on tape systems, but today other systems are available. Consequently the problem of choosing "systems within systems" is compounding the non-standardization problem. Each basic system format (e.g., tape, disc, or film) has several distinct system types which are incompatible. (See Tables 12 and 13.)

The video-tape format offers the greatest choice, with systems using quarter-inch, half-inch, three-quarter-inch, and one-inch tape sizes. This wide choice is the result of the advantages and features of each system. Cost is one important factor. In general, the larger the tape, the higher the cost of equipment and software. Other factors may be equally important. The availability of 60 minutes of playing time with stereo sound makes the ¾-inch system advantageous, but for many circumstances the 30-minute playing time and mono sound of ½-inch may be adequate. The portability of the unit is significant. Smaller size tape machines are more easily moved and are, therefore, more suitable for use in remote locations. The availability of a color camera also may be a decisive factor in choosing a particular system. Basically, the advantages of video tape over competing technology include: instant viewing, recording capability, and the ability to re-use cassettes.

Video discs do not offer the recording capability of video tape and, at present, are restricted to playback only. Three basic types of video discs are now available. (See Table 13.) The mechanical type (e.g., Teldec) operates on the same principle as audio discs, using a pressure stylus to derive information from modulated grooves. The advantage of this system is its simple technology, but its disadvantages include shorter playing time and higher disc wear. The second type is the optical disc (e.g., MCA or Philips), which uses a low-powered laser to pick up information without touching the record. This involves a more complicated technology, but it results in longer playing time and eliminates wear. The last possibility is the magnetic or electrical disc, which uses the same concept as magnetic video tape.

Video players using the film format are basically of two types, Electronic Video Recording (EVR) and Super 8. EVR, developed by CBS, is a miniature film, playback-only system. CBS has dropped plans for using the system in this country but is currently marketing EVR in Japan, Canada, Australia, and other countries.

Table 12
CHARACTERISTICS OF VIDEOPLAYER TAPE SYSTEMS

Type	Manufacturer	Recorder/ Player Cost	Maximum Playing Time	Maximum Cost per Blank Tape	Tape Rental Cost	Date Available
¼ inch	Akai	$6,000 w/camera	...	...	...	By or before 1973
½ inch Cartrivision	Cartridge Television, Inc.	$1,250	112 min.	...	$5–$7	(Defunct)
Electronic Industries of Japan (E.I.A.J.)	Panasonic Victor Co. of Japan	$1,495–	60 min.	$35.00	...	By or before 1973
	Sanyo	...	...	...	...	...
	Sharp	$1,300	...	...	...	...
E.I.A.J. Portapack	Victor Co. of Japan	$6,950	...	...	...	1974
Video Cassette Recorder	Philips	$1,295	90 min.	$32.50	...	By or before 1973
¾ inch U-Matic	Sony Victor Co. of Japan Wollensak Panasonic Concord MCA	$1,525	60 min.	...	...	By or before 1973
Portable U-Matic Mag. Tape	Sony RCA Bell & Howell	$3,000 $ 800 ...	20 min. 60 min. ...	$30.00	...	By or before 1973 1975 ...
1 inch	IVC	...	...	...	...	...

Source: Videoplayer Comparison Charts, *Proceedings of Video Publishing Year III* (White Plains, N.Y.: Knowledge Industry Publications, 1972), pp. 135–38; 1973 issues *Merchandising Week, Billboard, Radio-Electronics.*

Table 13

CHARACTERISTICS OF DISC AND FILM VIDEOPLAYER SYSTEMS

Technology	Type	Company	Player Cost	Disc Size	Playing Time	Disc Cost
			Disc Systems			
Mechanical	Ted	Teldec	$350	8"	20–25 min.	$4.20–$10.40
		Sanyo				
Electrical	MDR	Bogen	$400	12"	12 min.	$4.00
Optical	Discovision	MCA	$400	12"	40 min.	$2.00–$10.00
"	VLP	Philips	...	12"	30–45 min.	$10.40
"	...	Zenith	...	...	...	...
"	...	RCA	...	...	...	...
"	...	Thompson/CSF	$700	...	20–25 min.	...
"	IOM	Yo Metrics	$300	...	60 min.	$5.00
			Film Systems			
...	EVR	Hitachi	...	...	60 min.	...
...	Super 8	Kodak	$1,095 (Camera $500)	...	(50, 100, 200, or 400 ft. reels)	...

Source: Videoplayer Comparison Charts, *Proceedings of Video Publishing Year III*, Sept. 1972, pp. 135–38, but were updated using 1973 information in articles from *Merchandising Week, Billboard,* and *Radio-Electronics.*

Super 8 video-players, developed by Kodak, provide playback on an ordinary television set of films taken by any Super 8 camera. This means that any films made for Super 8 optical projectors can be used in the new video-players. With the recent addition of Super 8 cameras with magnetic sound and a processor that develops film in less than 15 minutes, Super 8 is competitive with tape systems. Super 8 film's biggest advantage is its versatility—it can be used in either a video player or a projector—but it also has the disadvantage of being usable only once.

The Video Cassette Market[2]

Video cassettes are marketed primarily for institutional, industrial, and medical use. The world-wide market for cassettes, which is growing rapidly, is dominated by two companies. In Europe, Philips controls 70–90 percent of the market with its VCR. Philips total production of the ½-inch units was expected to reach 75,000 by the end of 1973. Only a few thousand of these units are for the U.S. market, however. Sony controls the market in the United States and Japan. Its total production of ¾-inch U-Matics during 1973 was 100,000 compared to 50,000 in 1972, with 70 percent of these units for sales in the United States and Canada. Sony is clearly the leader in the United States selling over 50,000 units when only 10,000 units sold for all other formats.

In the United States various businesses and schools are beginning "video cassette networks" to distribute programs throughout an organization. One of the best examples is the Arthur Young & Company accounting firm, which has equipped fifty of its sixty U.S. offices with video recorders. Each office receives two to three programs monthly from the home office for the purposes of training and corporate communication. Among many other businesses using similar cassette networks are Manufacturers Hanover Trust in New York, Alexander and Alexander insurance brokers, Xerox Corporation, and New York Life Insurance.

The educational use of video cassettes in the United States is still quite limited—1973 estimates of penetration ranged from 6 to 10 percent of the market—lack of funds, skepticism about new technology, and copyright problems all being factors which tend to limit use. Most of the school systems using video cassettes have few player/recorders per building and do most of their own programming. For example in 1973 the Kenosha, Wisconsin, system had seven players; Abilene, Texas, had two; Denver, Colorado, had nineteen players; and the state of South Carolina, which had 806 public schools, had 361 players. Institutions of higher education use video cassettes to a slightly greater extent, particularly medical schools.

Obviously, the market for video cassettes will grow more rapidly with the development of home video cassette units. Present penetration of the home is difficult to determine, because most manufacturers are unwilling to give sales information. Sony, however, has revealed that only 10 percent of its production has been sold to individual consumers, while 90 percent of its sales are distributed equally between the educational and institutional markets. The percentage of home sales is significant, though, because Sony does not have a unit designed for the home and its present U-Matic unit costs almost $1,600. Cartrivision, the only company so far to market a video cassette designed to be a unit, went out of business in 1973 as a result of financial problems. Thus, the market is largely untested.

Widespread home use of video cassettes is limited by two factors, cost and compatibility. Industry observers generally contend that a price of about $400 for a television add-on unit will provide the necessary quality at a marketable price. A higher price would limit consumer demand to an unprofitable level. Full volume prices are expected to approach $400 in all three formats—tape, disc, and film.

Software costs are more difficult to predict. Much of the cost of prerecorded cassettes will be in talent fees or performer rights. Thus, the software market will be dependent on a large volume, which means some standardization will be necessary and the market for either tapes or discs ought not be fractionalized by competing technologies.

There will probably be standardization within systems and co-existence between film/disc and tape, a situation similar to that in the audio market in which tape is used for recording by hobbyists and discs are used only for playback. Significantly, the experience of Cartridge Television, Incorporated, showed that recording capability was the strongest selling point for Cartrivision. But, if Zenith or RCA offers a disc system with the cost of discs substantially lower than the cost of prerecorded tapes, discs and tape will each appeal to different markets.

Rental of video cassettes and use of cassettes on CATV may provide alternatives to the high cost of software. Video cassette rental is important in that most programming will not be used as repeatedly as audio records or tapes. In addition, tape distributors can record over video tapes that are not successful, thereby cutting inventory. The practicability of rental is dependent on solving copyright problems, however.

The Future

According to the 1973 Creative Strategies report *The Video-player Industry*, world-wide sales of video cassette units will have increased from 63,000 in 1972 to 282,000 by 1977. Hardware revenues will have increased from $56 million to $178 million in the same period, and software revenues are expected to grow to $78 million by 1977 from $8.2 million in 1972—i.e, at a compound annual rate of 57.2 percent.

In the United States alone, annual sales to consumers are expected to jump from 7,000 units in 1972 to 100,000 in 1977. Thus, penetration of less than 1 percent of television households is forecast, with the real growth of the home market coming sometime after 1977. The U.S. industrial-educational market is expected to increase from $27 million in 1972 to nearly $60 million in 1977 at a compound annual rate of 16.7 percent.

Significantly, the growth of the video cassette industry will require from $50 million to $150 million of investment capital, which should restrict industry leadership to present leaders or to large companies.[3]

NOTES

1. Materials for this section were compiled from a variety of sources. Especially helpful were the following: *Video Cassette Systems Industry* (Palo Alto, Calif.: Creative Strategies, Inc., June 1971); "VTR's—Many Different Systems," *Radio Electronics*, June 1973, pp. 39–41; *Proceedings of Video Publishing Year III, Year IV*, Sept. 1972, Sept. 1973, sponsored by Knowledge Industry Publications, Inc., White Plains, N.Y.

2. Information compiled from *Billboard*, Oct. 16, 1973; *Variety*, Oct. 10, 1973; *Merchandising Week*, June 11, 1973, and Sept. 3, 1973.

3. *The Videoplayer Industry* (Palo Alto, Calif.: Creative Strategies, Inc., 1973), pp. 3–5.

VII

SUMMARY AND CONCLUSIONS

The outlook for future growth in CATV depends greatly on the degree of success CATV operators experience in increasing subscriber penetration in the top 100 television markets. Partial relaxation of FCC regulations in 1972 improved the climate for entry into or expansion of CATV service in these markets. If this opportunity is to be exploited, however, several key problems must be overcome.

Heavy investment is required to build or expand CATV systems in major metropolitan markets. Providing the capital necessary for this expansion is a serious challenge, especially in view of high interest rates being charged on CATV business loans. Even if these rates were to decline to a level which would encourage CATV operators to consider borrowing, potential lenders would be especially concerned about the extent to which CATV revenues might be expected to grow in the top 100 markets, since this would indicate whether the prospects of profits were favorable enough to justify the risk involved in making long-term loans to CATV systems.

If revenues are to show satisfactory growth, CATV operators must develop and market a package of services that will (1) attract new subscribers to the CATV system, and (2) cause existing subscribers to buy added services. This function must be handled in a manner that will provide CATV with a competitive edge.

Strategy for the Top Fifty-Five Markets

In choosing the package of services to offer subscribers, CATV operators may find it useful to divide the top 100 markets into two groups according to the amount and type of over-the-air television service available. The more promising category should include 55

126.

of the 100 major television markets in which all three television networks, as well as an educational station, are available to viewers in good quality and over the air, but where no independent television service exists. Observers believe it is here that CATV has the greater potential for building penetration. Possible approaches CATV operators should explore to gain subscribers include the following:

1. *Import those programs offered by distant authorized independent television stations which have strong audience appeal.* This may prove to be the principal attraction that stimulates consumers to subscribe to CATV.

2. *Offer premium entertainment (pay television) on which the viewer may enjoy programs without commercial interruptions.* A substantial portion of television viewers are annoyed by the number of commercial interruptions they must tolerate to view movies on over-the-air programs. The early start of reruns, which has characterized over-the-air television in recent seasons, also reduces viewers' options when seeking attractive programs during their customary viewing hours. Availability of outstanding features on pay cable television might be welcomed by many television viewers when nothing new and interesting is coming via the networks.

   The number of firms available to provide pay-cable services for the system operator facilitates adding this extra service, but the cable operator must still decide which pay-cable supplier to contract with and how much of the work involved in offering premium entertainment he wishes to do himself. In choosing between alternative suppliers the operator should decide: (a) what type of premium entertainment will appeal most strongly to the television-viewing audience in the area served by the CATV system; (b) which premium entertainment supplier offers the highest quality entertainment from the viewpoint of the potential television audience; (c) which supplier has the best record of furnishing a converter that gives trouble-free operation; (d) which alternative promises to keep the cost of pay-cable service to the subscriber low enough to make it attractive when compared with hiring a baby-sitter and going out to the theater or sports arena; (e) which pay-cable system makes it easier for the subscriber to purchase the premium entertainment; (f) which supplier has a reputation for providing effective marketing services at a reasonable cost during both the introduction of pay-cable service and in subsequent periods of follow-up promotion; (g) how revenue-sharing arrangements, provision of services to facilitate the pay-cable operation, and the coverage of associated expenses compare, as spelled out in the contracts offered by alternative suppliers. Clearly the operator's goal is to choose the pay-cable system that promises to bring the maximum profit.

3. *Offer other local programs which may also be significant attractions for television viewers*—e.g., local sports events, religious presentations, other special features that may appeal to particular audiences.

4. *Originate programs having a special appeal to a local audience* (closely related to the above). If really good program ideas are to be produced, the CATV system must invest in them. The possibility of developing cable program networks through which a popular show can be made available to CATV systems in other areas is one way to recover programming costs and benefit from the audience appeal of successful presentations.

A survey which checked on the activity of large multiple systems operators (MSOs) in program origination and distribution produced nine responses from operators controlling from five to 140 individual CATV systems and serving from 20,000 to 800,000 subscribers. Together the respondents controlled an aggregate of 426 individual CATV systems with a total of over 2 million subscribers. Seven out of the nine produced programs for their own systems. Average weekly production ranged from 20 to 364 hours with a mean of 84.8. (If the largest figure is eliminated to avoid distortion of the mean, the average works out to 38.3 hours per week.) Five of the nine MSOs indicated that individual CATV systems produced programs for other systems of the company. The individual systems who produced, however, represented only 55 out of a total of 273 systems. Of the nine MSOs surveyed, four had interconnection capabilities and a total of six intend to provide them in the future. Projected programming plans include networking by satellite (seven out of nine) and cooperation with other MSOs (five out of nine).

By way of comparison, in 1973, of the nation's 3,000 cable systems, 585 originated programming. These systems served 1,363 communities and reached nearly four million subscribers. The average amount of origination per system was 21 hours per week (as opposed to an average of 38 hours per week for the MSOs mentioned above).

Other interesting possibilities for making good local origination programs widely available are interconnection of CATA systems via microwave or by satellite. Such suggestions are currently being developed and may soon constitute a practicable approach to providing entertainment that will attract subscribers to CATV systems.

An additional development that has much promise in regard to good programming is the organizing of regional cable system networks around outstanding independent over-the-air broadcasting stations. To check on this development, in the fall of 1973 a questionnaire was sent to 30 (of the 90) major independent stations in various parts of the United States. Returns were received from fourteen. The results indicate a wide range in the number of CATV systems carrying the signals of the various independent stations: i.e., Station No. 7 was carried by 50 CATV systems with 150,000 subscribers; Station No. 1 was carried by 204 systems with 600,000 subscribers. The responding independent stations' signals were carried by an average of 109 CATV systems. These systems served a mean of 339,739 subscribers. (See Chapter 5, Appendix A.) Clearly there is a potential here for the development of regional program networks organized around such independent stations. Such a development would be mutually beneficial to the independent over-the-air TV stations and CATV system operators.

Strategy for the Second Group of Markets

The marketing strategy for the second group of major markets, the remaining and somewhat less promising 45, presents a more challenging problem. Here television viewers have readily available the three networks, one, two, or three independent stations, plus one or two educational outlets—all capable of being received clearly, often with rabbit ears. Obviously, the CATV system has little to offer in terms of either improved clarity of reception or diversity of programming. For viewers in such markets CATV service is an added, optional luxury. Those who do subscribe are more likely to drop the CATV service if it proves disappointing than are subscribers in communities where program options are fewer or where over-the-air reception is poor. What alternatives should the CATV operator consider in such markets?

First, experimentation with premium (pay) cable television is clearly in order. Even in these markets, television viewers may respond favorably to full-length movies, musicals, or sporting events that are presented without commercial interruptions. Here, too, disgust with the early start of reruns on over-the-air television may make a portion of the viewers receptive to CATV program offerings.

Second, offering various automated program services via CATV may round out a program package that will attract some viewers. Services such as news, weather, stock ticker, time, and music may be considered as possibilities.

Third, clearly substantial investment in originating local programs, participation in cable program networks, and similar devices would merit top priority consideration in planning a market strategy. If CATV can develop programs that have special appeal to local viewers—features not available on over-the-air network and independent stations—then the CATV operator will have a competitive edge to exploit in attracting subscribers and in enticing them to buy more than the basic service offered via the cable system.

Fourth, since FCC regulations require all systems constructed after March 1972 to provide two-way transmission capability, cable operators in the top 100 markets may also turn to new services that draw on this technology. One service seriously considered by cable operators is that of home security systems. This might include automatic warning systems for fire and burglary that would alert not only the homeowner, but also the fire or police department. Studies are needed to determine the potential demand for such security systems as well as the costs involved in providing these services.

Electronic shopping service is another application of the two-way interactive capability that is being explored by a number of cable operators. A related development attracting considerable interest is experimentation with electronic banking. While these two interactive services may not become available immediately, they hold long-run promise. CATV operators should follow closely experiments involving such two-way services.

For the longer view, it is interesting to note the experts' predictions as to the introduction of various two-way services and the anticipated growth in the demand for these services by 1990. In a study using the Delphi technique these experts forecast the potential market for thirty possible two-way services. The five individual services with the most favorably predicted prospects include (in order and in millions of dollars): (1) plays and movies from video library, $2,829; (2) computer-aided school instruction, $2,047; (3) cashless-society transactions, $1,810; (4) person-to-person (paid work at home), $1,713; (5) computer tutor, $1,414. Shopping transactions were estimated at $859 million, which placed them thirteenth on the list. Grocery price lists, bringing in forecast revenues of $566 million, were fifteenth. Special sales information was estimated at $354 million (sixteenth). Providing a consumer advisory service was also expected to yield $354 million.

Since some of the services that may be rendered by CATV systems may also be offered by competing alternative approaches, the development and use of the video cassette has been examined briefly. As of 1974 several factors (lack of standardization and compatibility, high cost, programming shortages, copyright problems, etc.) had kept the video cassette from achieving widespread use. If, however, these limitations are overcome in the next decade, as observers expect, widespread use of video cassettes will affect television and CATV programming, motion picture theatres, school curricula, and other facets of life. The home market for video cassettes is largely untested. At present, the major market for video cassettes is for institutional, industrial, or medical use. The extent to which the home market has been penetrated is difficult to determine since most companies producing video cassette equipment are unwilling to give sales information, but, as noted in Chapter 6, Sony has revealed that 10 percent of its production has been sold to consumers (educational and institutional markets accounting for 45 percent apiece). Sony does not have a unit especially designed for the home, however, and their present U-Matics are priced at almost $1,600. Cartrivision, the only company to market a home video cassette unit so far, went out of business in 1973 as a result of financial problems.

Use of video cassettes in the home is not yet widespread because of cost and problems with compatibility. Industry observers generally contend that a target price of about $400 is necessary for an add-on unit to be successful. Full volume prices are expected to approach that figure in all three formats—tape, disc, and film.

Although it is estimated that world-wide sales of video cassette units will increase from 63,000 in 1972 to 282,000 in 1977 and that in the United States alone annual sales to consumers will jump from 7,000 units in 1972 to 100,000 in 1977, that means that penetration will still be less than 1 percent of television households by 1977, with the real growth of the home market coming thereafter. It therefore appears that CATV operators are not likely to meet significant competition from video cassettes in the home market in the immediate future. In the long run, however, this picture might change, if video cassette hardware and software prices are reduced and problems of standardization and compatibility are resolved. Developments in this field should therefore be followed closely.

From the foregoing it is clear that the CATV operator needs to be consumer oriented in selecting a marketing strategy to build penetration in the top 100 television markets. A growing number of cable systems are offering premium entertainment (pay cable television) as a means of increasing revenue and attracting new subscribers. Analysis of potential markets for two-way CATV services—such as home security, checkless banking, home computing, and electronic shopping, among others—is badly needed but has not yet been undertaken extensively. Decision making on what two-way services to offer would be greatly assisted by investments in experiments in which certain promising services were offered' in a limited area. Data could then be collected on consumer demand for such services at several possible prices as well as on the cost of providing them under operating conditions.

Thus, the CATV operator must prepare to move aggressively to execute carefully thought-out marketing plans as the environment for cable television improves.